FACING UP

Statue of Tycho Brahe, at the site of his observatory
on the island of Hven

FACING UP

Science and Its
Cultural Adversaries

———

STEVEN WEINBERG

Harvard University Press

Cambridge, Massachusetts

London, England

Copyright © 2001 by Steven Weinberg
All rights reserved
Printed in the United States of America
Second printing, 2003

First Harvard University Press paperback edition, 2003

Library of Congress Cataloging-in Publication Data

Weinberg, Steven, 1933–
Facing up : science and its cultural adversaries / Steven Weinberg.
p. cm.
Includes bibliographical references and index.
ISBN 0-674-00647-X (cloth)
ISBN 0-674-01120-1 (paper)
1. Science.   I. Title.

Q171 .W419   2001
500—dc21   2001024219

To Louise and Elizabeth

Contents

Preface

In the first year that we were married, my wife and I lived in an attic apartment on the coast road running along the Danish side of the strait between Denmark and Sweden. From our living room window we could see a little island in the strait, near the Swedish side. Our landlord told us that the island was named Hven. After returning to America we learned that Hven was the island where in the 1570s the Danish astronomer Tycho Brahe had built his observatory, Uraniborg. Without a telescope, but using giant naked-eye instruments anchored to massive foundations, Brahe made historic measurements of angles between stars and planets. It was partly the great precision of these measurements that allowed Tycho's successor Johannes Kepler to infer that the planets move on ellipses rather than circles, a result that later was of crucial importance to Isaac Newton in developing his theory of gravitation.

Years after, on a visit to Copenhagen one summer, my wife and I and our daughter finally had a chance to visit Hven. We took the ferry across the strait to the island, and drove out in a taxi to the site of Uraniborg. All around was farm land, with nothing left of the observatory but its impressive foundations. Above ground there was only a granite statue of Brahe, carved in 1936 by the Danish sculptor Ivar Johnsson. A photo of the statue appears facing the title page of this book. As can be seen, the statue shows Brahe in a posture appropriate for an astronomer, *facing up.*

That is only part of the reason for my choice of *Facing Up* for the title of this collection of essays. (Brahe is not one of my special heroes; he rejected the idea of Copernicus that the earth goes around the sun, and he was a rotten landlord.) The researches of

Brahe, Kepler, Newton, and their successors have presented us with a cold view of the world. As far as we have been able to discover the laws of nature, they are impersonal, with no hint of a divine plan or any special status for human beings. In one way or another, each of the essays in this collection struggles with the necessity of facing up to these discoveries. They express a viewpoint that is rationalist, reductionist, realist, and devoutly secular. Facing up is, after all, the posture opposite to that of prayer.

Most of my working life has been devoted to research in physics and astronomy. My papers were published in *The Physical Review* and other scientific journals, and I did not expect to do much writing outside the technical literature of physics and astronomy. Then in the 1980s I started to speak and write in defense of spending on research in science, and in particular on the Superconducting Super Collider, a large and controversial facility for research in elementary particle physics. I found that I had a taste for controversy, and I began to accept invitations to write and speak on wider issues—on the follies that I found in the attitudes toward science of many sociologists, philosophers, and cultural critics, and on the ancient tension between science and religion. Even so, only a few of these essays were initiated by me. Something like a chain reaction took place—articles when published led to invitations to write other articles or to give talks that I then wrote up as articles. But I would not have written these essays if I had not enjoyed it so much.

The essays in this collection are presented here in chronological order, and pretty much as they were first published over the past fifteen years, with just a little editing to clarify some points and mitigate repetitions. I have added new introductions to all the articles to explain how the articles came to be written and bring them up to date where necessary.

I am grateful to Michael Fisher of Harvard University Press for suggesting the publication of a collection of my articles, and for his good advice. Thanks are due to Owen Gingerich for providing the photo of the statue of Brahe, and to Nancy Clemente for her sensitive and intelligent editing. I thank Terry Riley for finding countless books and articles, and Jan Duffy for many helps. For suggestions that I think greatly improved these articles I also owe thanks to many friends and to the editors of the periodicals in which the articles originally appeared, especially Robert Silvers of *The New York Review of Books.*

Austin, Texas
January 2001

FACING UP

1

Science as a Liberal Art

I like college commencements. The ceremonies are held in attractive settings, usually at the nicest time of the year, and, best of all, you are thrown together with interesting people whom you would not otherwise have met. The 1985 commencement at Washington College was no exception; Washington is a lovely old liberal arts college on the Eastern Shore of Maryland, and my wife and I had the pleasure, at a small dinner party the night before the commencement ceremony, of listening to Isaac Stern talking about the emotional impact of music.

The down side was that I had to give the commencement address. This is hard enough if you are an Isaac Stern, but I couldn't imagine anything in which undergraduates on the day of their graduation would be less interested than a talk by a theoretical physicist of whom they had never heard. I decided not to worry about it, and just talk about some things that were on my mind at the time. One of them was a subject to which I frequently return in the pages of this book, the effect of progress in science on the human spirit.

I also touched on another topic that was in the news then and is again now: the proposal that the United States should build a ballistic missile defense system. I was one of many scientists who had fought this proposal from the 1960s on, not only because of its technical problems but also because I thought that building an antimissile defense would lead the Soviet Union to increase its missile forces. Since 1985 the Soviet Union has collapsed, and some admirers of Ronald Reagan have claimed that it was the failed effort of the Soviets to keep up with American antimissile technology that led to the economic breakdown that ended Soviet communism. Nevertheless, I still think that I and my colleagues had been right to oppose deploying an antimissile system. For one thing, whatever alarm Soviet leaders felt when the Strategic Defense Initiative was first proposed, it is hard to believe that the prospect of an American antimissile program remained terribly frightening to the Soviets, when we manifestly were not building such a system, or even deploying the sort of antimissile system allowed by the 1972 arms control treaty. Public statements by the Reagan administration kept reassuring the Soviets that our Strategic Defense Initiative was not intended to threaten their strategic deterrent.

But suppose that Reagan and his advisers knew that the Soviets would feel threatened anyway, and would be led thereby to disastrous overspending on military programs, as some now claim. If this were true, then whatever its success, we were playing an incredibly dangerous game. Instead of overspending, the Soviets might have inexpensively preserved their deterrent by putting their forces on a hair-trigger "launch on warning" status, in which a radar or computer malfunction could start World War III. Not only is this a risk that we should never have taken—it is a risk that the American people were told that they were not taking. Reagan's admirers can't have it both ways; they can't deny that the Strategic Defense Initiative threatened the Soviet nuclear deterrent, and at the same time give it credit for wrecking the Soviet economy. I think that in fact our government was neither so diabolically clever nor so reckless, and that want of good judgment provides an adequate explanation for both the Soviet economic collapse and the Strategic Defense Initiative.

———

In discussing with President Cater[1] by phone what sort of commencement talk might be appropriate here today, the idea occurred to both of us that, since Washington College is setting out on a major renovation of its science facilities, and since I am a sci-

1. The late Douglas Cater, then president of Washington College. [Added note.]

entist, I might speak on the place of science education in small liberal arts colleges. But as soon as I hung up the phone, my heart sank. Many of you today are saying your goodbyes to college. I am afraid that for you to have to listen to me talk about education is much like passengers on a ship just coming into port after a long voyage having to listen to a sailor lecture on the principles of navigation.

But then I warmed to the task. Most of my life has been spent in studying or working in a different sort of educational institute, the large research university. The research university is a peculiar sort of institution that began in Germany in the nineteenth century, and was first transplanted to the United States not far from here, at Johns Hopkins, about a century ago. Our universities are marvelous places for faculty members and graduate students to do research, and as such they have been tremendously important to our country. I am convinced that without great research universities we in the United States would have to support ourselves by growing soybeans and showing the Grand Canyon to tourists from Germany and Japan.

But research universities are generally not institutions that focus on the role of science education. I don't say that no one in these universities cares about education, but it is research and not education that drives our most important decisions. After over twenty years of faculty meetings, I can say that I've never seen any physicist hired because he or she was a good teacher rather than a good researcher. But thank heaven for the variety in America! At small liberal arts colleges like Washington there is an intensity of concern about education that is rare at research universities.

Let me put aside right away the topic of science education for future scientists. In doing so, I don't for a minute mean to imply that small liberal arts colleges have no role here. In fact, they seem to do about as well in preparing future scientists for graduate school as do the large research universities. I just don't think that the undergraduate training of future scientists raises any deep questions—most of these students seem to get the important part of their education from self-propelled reading rather than from courses anyway. I want to talk here about the role of science in the education of undergraduates who have no interest in becoming scientists themselves.

The most frequently heard rationale for science in a liberal arts curriculum is that it is needed to help us understand the technological background of modern society. Maybe so, but in my view this is the least important aspect of science education. Take a current issue that has a lot to do with applied science, the administration's Strategic Defense Initiative, or "star wars defense." Now I happen to believe that, although it is worthwhile to do some quiet research in this area, as we've in fact been doing for decades (I've done some myself), the scale and publicity of the current program are terribly foolish. This isn't at all because I have some special knowledge about scientific matters like x-ray lasers; in fact, I think that it probably would be possible, at great expense, to build a satellite-based x-ray laser that would, under favorable circumstances, knock out a few missiles as they rose above the atmosphere. My worry about star wars arises from a general sense of skepticism about the motives of the policymakers who claim that this sort of system can work effectively in an unfavorable environment and, even more, from a sense of the likely reactions of both our adversaries (and our allies) when they see their deterrent threatened. I am much less worried that our leaders are not the sort of people who understand how lasers work than I am about the fact that they do not seem to be the sort of people who read history.

Much more important than the effect of science on our capabilities, it seems to me, is its effect on *ourselves*. Nothing in the last five hundred years has had so great an effect on the human spirit as the discoveries of modern science. Just think of the effect of the discoveries of Copernicus, Galileo, and Hubble in astronomy, and Darwin, Wallace, and Mendel in biology. We find that the earth on which we live is a speck of matter revolving around a commonplace star, one of billions in a galaxy of stars, which itself is only one of trillions of galaxies. Even more chilling, we ourselves are the end result of a vast sequence of breedings and eatings, the same process that has also produced the clam and the cactus.

I can't tell anyone what sort of spiritual conclusions they should draw from all this. Giordano Bruno wrote that when he learned that the earth was just a small part of the universe, he felt that he could breathe more freely. On the other hand, some of the old magic has gone out of our view of the role of humanity in the universe, its place being taken by what Matthew Arnold called the "note of sadness."

You know, our fundamentalist friends dislike the teaching of evolution in schools because of the effect they feel it has on our view of our own special importance, while liberals insist that scientific and spiritual matters can be kept in separate compartments. On this point, I tend to agree with the fundamentalists, though I come to opposite conclusions about teaching evolution because I am convinced it's true. The human race has had to grow up a good deal in the last five hundred years to confront the fact that we just don't count for much in the grand scheme of things, and the teaching of science as a liberal art helps each of us to grow up as an individual.

Even more important than the specific discoveries of science in astronomy and biology has been the discovery of science itself. Although we ourselves don't play a large role in it, there is a grand order: the tides and planets and stars move according to the same physical laws, and the laws that govern lightning are the same as those that control "the force that through the green fuse drives the flower." It's an old idea that the laws that rule us here on earth are the same as those obeyed by what we see in the sky. In George Meredith's poem *Lucifer by Starlight*, the archfiend rebels, and flies about doing all sorts of nastinesses, until

> He reach'd a middle height, and at the stars,
> Which are the brain of heaven, he look'd, and sank.
> Around the ancient track march'd, rank on rank.
> The army of unalterable law.

I fear that I may have been talking about science in such exaltedly existentialist tones that I have given the impression that science education is the moral equivalent of tearing up books of fairy tales while standing in a cold shower. Fortunately, there is one aspect of science that makes it attractive as part of a humanistic education. Scientific research is done not by gray angels, but by human beings. I was charmed to read recently in an issue of *ISIS* that after Galileo discovered the moons of Jupiter and named them the Medicean stars to flatter the Grand Duke of Tuscany, he wrote to the duke to reassure him that there were no more new stars to be discovered that might be named after someone else.

It's not just that scientists have their normal share of human roguishness. The scientific enterprise at its best depends on very human prejudices and preconceptions. I know that I did some of my own best work because I had certain preconceptions about the way forces ought to work, and ignored experimental evidence to the contrary, and I did not succeed in taking the next step in this work because I was prejudiced against certain mathematical methods. It's not an atypical story. I know of no better way of teaching science to undergraduates than through its history. Science is, after all, a part of the history of humanity, and not, I think, the least interesting part.

2

Newtonianism, Reductionism, and the Art of Congressional Testimony

Nineteen eighty-seven was the three hundredth anniversary of the publication of Isaac Newton's *Principia*, the great book in which Newton laid out his theory of motion and gravitation. It was natural for this anniversary to be celebrated at the University of Cambridge, where Newton had studied and later held the Lucasian Chair of Mathematics. As it happened, our daughter was spending the 1986–87 academic year doing research in Cambridge, and I had let it be known to all my friends there that during 1986–87 I would gratefully accept any invitation to visit Cambridge. I was duly invited to speak at the *Principia* celebration in June 1987, and this talk, later published in *Nature*, was the result.

At around this time I was spending a good deal of time arguing for the construction of a very large instrument for research in the physics of elementary particles, the Superconducting Super Collider. (This wasn't Texas patriotism; at the time the decision had not yet been made to locate the Super Collider in Texas.) It seemed to me that, along with all the practical, political, and economic issues raised by the Super Collider project, there was also a philosophical issue. The kind of physics research that is done in elementary particle laboratories derives much of its importance from a reductionist world view: we seek those fundamental principles from which all other scientific principles may in principle be derived. Newton may have been the first to express this aim, and he certainly did more than any other scientist to show how it could be realized, so reductionism seemed like a natural topic for my talk. Parts of the talk were later recycled and expanded in my 1992 book *Dreams of a Final Theory*. The Superconducting Super Collider project was canceled by Congress in 1993, but work continues on a similar though smaller accelerator at CERN, near Geneva. And the issues surrounding the support of research in the pure sciences will be with us for some time to come.

One correction is needed. In this talk I quoted Lord Kelvin saying around 1900 that "there is nothing new to be discovered in physics now. All that remains is more and more precise measurement." I had seen this quote in a copy of a transparency used in a physics conference talk given by a good friend. When I was writing *Dreams of a Final Theory* I tried to find the quote by going through Kelvin's biography and his selected works, but without success. I finally asked my friend about it, and he told me that this was only his recollection of a quote that he had heard attributed to Kelvin many years earlier. At present I have no idea what Kelvin actually said about there being nothing new to be discovered in physics, or if indeed he had anything at all to say about it.

My talk this afternoon will be about the philosophy of science, rather than about science itself. This is somewhat uncharacteristic for me, and, I suppose, for working scientists in general. I've heard the remark (although I forget the source) that the philosophy of science is just about as useful to scientists as ornithology is to birds.

However, at just this time a question has arisen in the United States that will affect the direction of physics research until well into the twenty-first century, and that I think hinges very largely on a philosophical issue. On January 30 of this year the present administration in Washington announced that it had decided to go ahead with the construction of a large new accelerator for elementary particle physics, the Superconducting Super Collider, or SSC for short. "Large" in this case means that its circumference would be about 53 miles. The circumference is determined by the necessity of accelerating protons to energies of 20 trillion electron volts. Within this ring there would travel two counter-rotating beams of protons that would slam into each other at a number of intersection regions. The intensity of the beams is designed to be such that one would have a rate of about one per second for collisions that result in interesting processes. All of these design parameters lead to a bottom-line parameter: the cost in 1986 dollars is estimated to be $4,400 million.

The chief reason for wanting to go ahead with this accelerator is that it would open up a new realm of high energy that we have not yet been able to study. Just as when astronomers start to study the sky at a new wavelength or when solid state physicists go down another factor of ten in temperature they often make dramatic new discoveries, when particle accelerators go up a factor of ten in energy we generally discover exciting new physics. This has often been the rationale for new accelerators. Occasionally, one can also point to specific discoveries that can be anticipated from a particular new accelerator. One example is provided by the accelerator built in Berkeley over thirty years ago, the Bevatron, which for the first time was capable of producing particles with masses of 1 billion electron volts. The Bevatron was designed to be able to produce antiprotons, and indeed it did so shortly after it went on line. That was not the only exciting thing done at that accelerator. Quite unexpected was the discovery of a vast forest of new types of particles, which led to a change in our conception of what we mean by an elementary particle. But in planning the Bevatron, it was nice to know in advance that at least one important discovery could be counted on.

The same is true now of the SSC. The SSC is designed so that it will discover a particle known as the Higgs particle provided that the Higgs particle is not too heavy. If the Higgs particle is too heavy, then the SSC will discover something else equally interesting.

Let me explain these remarks further. As you may have heard, there has been a certain measure of unification among the forces of nature. This unification entails the idea that the symmetry among the forces, specifically the weak nuclear force and the electromagnetic force, is spontaneously broken—that is, although this is a symmetry of the fundamental equations of the theory, it does not appear in observable physical states. It can't be spontaneously broken by the forces we know about, that is, the ordinary strong and weak nuclear forces and the electromagnetic force; therefore there must be a new force in nature which is responsible for the symmetry breaking, like the well-known force in a superconductor due to the transmission of sound waves between electrons. We don't know exactly what that new force is. The simplest picture of this force is that it is associated with a new kind of elementary particle, the Higgs particle.

Now, we are not sure that that is actually the correct picture of the mechanism for electroweak symmetry breaking, and we certainly do not know the mass of the Higgs particle. The SSC would be able to discover the Higgs particle if its mass is not greater than about 850 billion electron volts and, of course, if it exists. However, the SSC (to borrow a phrase from Michael Chanowitz)[1] is a no-lose proposition because if the Higgs particle does not exist, or is heavier than 850 billion volts, there would have to be strong interactions among the particles that transmit the weak nuclear force, which the SSC could also discover. These strong interactions would reveal the nature of the spontaneous symmetry breaking betweeen the weak and the electromagnetic interactions.

Now it remains for Congress to decide whether or not to authorize construction of the accelerator and to appropriate the money. Two committees of the two houses of the Congress, the Committee on Space, Science, and Technology of the House of Representatives and the Subcommittee on Energy Research and Development of the Senate Committee on Energy and Natural Resources, announced hearings on the SSC, both to begin on April 7 of this year.

1. Chanowitz, M. S., talk presented at the 23rd International Conference on High Energy Physics, Berkeley, California, July 16–23, 1986 (Lawrence Berkeley Laboratory Publication no. 21973).

In March, about a month before these hearings, I was asked to testify at them. I must admit that I found this more frightening than inviting. I had been active for some time in working for the building of the SSC, and all this time it had been a nightmare of mine that I would be called up before some tribunal and asked in a stern voice why it is worth $4.4 billion to find the Higgs particle. Also, I had testified in Congress only once before, and I did not consider myself a master of the art of congressional testimony.

The particle physicists of the United States are in fact quite united behind the idea that this is the right accelerator to build next. (As I said, its purpose is not limited to finding the Higgs particle, which is just one target, but rather it is to open up a new range of energies.) But there has been substantial opposition to the SSC from other physicists in the United States. I have read that this is perhaps the most divisive issue that has ever faced American physicists.[2] I believe that here in Britain there is a similar debate—not about building an SSC but about whether Britain should remain in CERN, the European physics research consortium, an issue on which I gather not all British scientists agree.

I knew at the hearings in Washington there would be two heavyweights who would be testifying vigorously against going ahead with the SSC. One would be Philip Anderson, known to everyone as among the leading physicists in the world working on properties of condensed matter such as superconductivity. Anderson has over many years opposed the large sums that are spent on high energy physics. Another to testify would be James Krumhansl, also a distinguished solid state physicist. He, as it happens, taught me physics when I was a freshman at Cornell, but in addition, and this I suspect counts for more, he is slated the year after next to be the president of the American Physical Society.

Both Anderson and Krumhansl I knew would oppose the SSC, and they would be making some arguments with which I really couldn't disagree. In particular, I expected that they would ar-

2. Dixon, B., *The Scientist* (June 15, 1987), p. 13.

gue that money spent on elementary particle physics, high energy physics, whatever you want to call it, is not as sure to yield immediate technological advances as the same money spent on condensed matter physics, and some other fields. I would have to agree with that (though I would put more emphasis on the benefits of unpredictable discoveries and spinoffs). I expected that they would also argue that elementary particle physics is not more intellectually profound than other areas of physics like, say, condensed matter physics. I would also agree with that. In fact, we've seen in the last few decades a continual trading back and forth of ideas between elementary particle physics and condensed matter physics. We partricle physicists learned about broken symmetry from condensed matter physicists; they learned about a mathematical method known as the renormalization group from us. And now we're all talking about a class of physical theories known as conformal quantum field theories (I don't now who learned that from whom). But it is clear that there's no lack of mathematical profundity in condensed matter physics as compared with elementary particle physics.

The case for spending large sums of money on elementary particle physics has to be made in a different way. It has to be at least in part based on the idea that particle physics (and here, parenthetically, I should say that under "particle physics" I include quantum field theory, General Relativity, and related areas of astrophysics and cosmology) is in *some* sense more fundamental than other areas of physics. This was denied more or less explicitly by Anderson and Krumhansl in their testimony and also by most of the opponents of the SSC. I didn't see how I could avoid this issue in making a case for the SSC. But it's a dangerous argument. It tends to irritate one's friends in other areas of science. Let me give an example, and here I will quote from myself because then I want to quote some comments on my own remarks.

In 1974, shortly after our present Standard Model of elementary particles was put into its final form with the success of the theory of strong nuclear forces known as quantum chromodynamics,

I wrote an article[3] for *Scientific American* called "Unified Theories of Elementary Particle Interactions." Just to get the article started I began it with some platitudes, as follows: "One of man's enduring hopes has been to find a few simple general laws that would explain why nature with all its seeming complexity and variety is the way it is. At the present moment the closest we can come to a unified view of nature is a description in terms of elementary particles and their mutual interactions." I really didn't intend to make any important point by this; it was just the sort of thing one says (as, for instance, Einstein: "The supreme test of the physicist is to arrive at those universal elementary laws from which the cosmos can be built up by pure deduction"). Then a decade later I was asked by the MIT Press to review a proposed book, a collection of articles by various scientists. In the manuscript I found an article[4] by a friend of mine at Harvard, Ernst Mayr, who is one of the most eminent evolutionary biologists of our times. I found that Mayr cited my remarks in the *Scientific American* article as "a horrible example of the way physicists think." He called me "an uncompromising reductionist."

Now, I strongly suspect that there is no real disagreement between Ernst Mayr and myself, and that in fact we are simply talking past each other, and we should try to understand how we agree rather than fight over this. I don't consider myself an uncompromising reductionist. I consider myself a compromising reductionist. I would like to try to formulate in what way elementary particle physics is more fundamental than other areas of physics, trying to narrow this down in such a way that we can all agree on it.

Let me first take up some of the things I don't mean. And here it is useful to look back at some more of Ernst Mayr's writing, because he is in fact the leading opponent of the reductionist tendency within biology, as well as in science in general. He wrote

3. Weinberg, S., *Scientific American* 231, 50 (1974).
4. Mayr, E., in *Evolution at a Crossroads*, ed. Depew, D. J., and Weber, B. H. (MIT Press, Cambridge, 1985).

a book[5] in 1982, *The Growth of Biological Thought,* that contains a well-known attack on reductionism, and so I looked at it to see what Mayr thought reductionism was, and whether or not I should consider myself, in his terms, a reductionist.

The first kind of reductionism that Mayr opposes is called by him "theory reductionism." As far as I can understand it, it's the notion that the other sciences will eventually lose their autonomy and all be absorbed into elementary particle physics; they will all be seen as just branches of elementary particle physics.

Now I certainly don't believe that. Even within physics itself, leaving aside biology, we certainly don't look forward to the extinction of thermodynamics and hydrodynamics as separate sciences; we don't even imagine that they are going to be reduced to molecular physics, much less to elementary particle physics. After all, even if you knew everything about water molecules and you had a computer good enough to follow how every molecule in a glass of water moved in space, all you would have would be a mountain of computer tape. How in that mountain of computer tape would you ever recognize the properties that interest you about the water, properties like vorticity, turbulence, entropy, and temperature?

There is in the philosophical literature a term, *emergence,* that is used to describe how, as one goes to higher and higher levels of organization, new concepts emerge that are needed to understand the behavior at that level. Anderson summarized this neatly in the title of an interesting article[6] in *Science* in 1972: "More Is Different."

Another kind of reductionism is called by Mayr "explanatory reductionism." As I understand it, it is the idea that progress at the smallest level, say the level of elementary particle physics, is needed to make progress in other sciences, like hydrodynamics, condensed matter physics, and so on.

5. Mayr, E., *The Growth of Biological Thought:* 58–66 (Harvard University Press, Cambridge, 1982).
6. Anderson, P., *Science* 177, 393 (1972).

I don't believe that either. I think we probably know all we need to know about elementary particle physics for the purposes of the solid state physicist, for instance, and the biologist. Mayr in his hook makes a point that surprised me (but I suppose it's true; he knows a lot more about this than I do), that even the discovery of DNA was not really of much value in the science of transmission genetics. Mayr writes, "To be sure the chemical nature of a number of black boxes in the classical genetic theory were filled in by the discovery of DNA, RNA, and others, but this did not affect in any way the nature of transmission genetics."

I don't disagree with any of this, but it seems to me that in their attacks on reductionism, Mayr, and also physicists like Anderson, Krumhansl, and others, are missing the point. In fact, we all do have a sense that there are different levels of fundamentalness. For instance, even Anderson[7] calls DNA the "secret of life." We do have a feeling that DNA is fundamental to biology. It's not that it's needed to explain transmission genetics, and it's certainly not needed to explain human behavior, but DNA is fundamental nonetheless. What is it then about the discovery of DNA that was fundamental to biology? And what is it about particle physics that is fundamental to everything?

Having spoken at length about what I don't mean, now I want to say what I do mean. But I'm not trying here to say anything new, that you don't all already know. What I'm trying to do is precisely the opposite: to identify what we can all agree on.

In all branches of science we try to discover generalizations about nature, and having discovered them we always ask why they are true. I don't mean why we believe they are true, but why they *are* true. Why is nature that way? When we answer this question the answer is always found partly in contingencies, that is, partly in just the nature of the problem that we pose, but partly in other generalizations. And so there is a sense of direction in science, that some generalizations are "explained" by others.

7. Anderson, P., letter to the *New York Times,* June 8, 1987.

To take an example relative to the tercentenary celebration of the *Principia:* Johannes Kepler made generalizations about planetary motion, Newton made generalizations about the force of gravity and the laws of mechanics. There is no doubt that historically Kepler came first and that Newton, and also Edmund Halley and Christopher Wren and others, derived the inverse square law of gravity from Kepler's laws. In formal logic, since Kepler's laws and Newton's laws are both true, either one can be said to imply the other. (After all, in symbolic logic the statement "A implies B" just means that it never happens that A is true and B isn't, but if in fact A and B are both true then you can say that A implies B and B implies A.)

Nevertheless, quite apart from formal logic, and quite apart from history, we intuitively understand that Newton's laws of motion and law of gravity are more fundamental than Kepler's laws of planetary motion. I don't know exactly what I mean by that; presumably it has something to do with the greater generality of Newton's laws, but about this also it's hard to be precise. But we all know what we mean when we say that Newton's laws "explain" Kepler's. We probably could use help from professional philosophers in formulating exactly what that statement means, but I do want to be clear that it is a statement about the way the universe is, not about the way physicists behave. In the same way, even though new concepts "emerge" when we deal with fluids or many-body systems, we understand perfectly well that hydrodynamics and thermodynamics are what they are because of the principles of microscopic physics. No one thinks that the phenomena of phase transitions and chaos (to take two examples quoted by Krumhansl) could have been understood on the basis of atomic physics without creative new scientific ideas, but does anyone doubt that real materials exhibit these phenomena because of the properties of the particles of which the materials are composed?

Another complication in trying to pin down the elusive concept of "explanation" is that very often the "explanations" are only in principle. If you know Newton's laws of motion and the inverse square law of gravity you can deduce Kepler's laws—that's not so hard. On the other hand, we also would say that chemical behavior, the way molecules behave chemically, is explained by quantum mechanics and Coulomb's law, but we don't really deduce chemical behavior for very complex molecules that way. We can for simple molecules; we can explain the way two hydrogen atoms interact to form a hydrogen molecule by solving the fundamental equation of quantum mechanics called Schrödinger's equation, and these methods can be extended to fairly large molecules, but we can't work out the chemical behavior of DNA by solving Schrödinger's equation. In this case we can at least fall back on the remark that although we don't in fact calculate the chemical behavior of such complicated molecules from quantum mechanics and Coulomb's law of electrical attraction, we could if we wanted to. We have an algorithm (the variational principle), which is capable of allowing us to calculate anything in chemistry as long as we have a big enough computer and are willing to wait long enough.

The meaning of "explanation" is even less clear in the case of nuclear behavior. No one knows how to calculate the spectrum of the iron nucleus, or the way the uranium nucleus behaves when fissioning, from quantum chromodynamics. We don't even have an algorithm; even with the biggest computer imaginable and all the computer time you wanted, we would not today know how to do such calculations. Nevertheless, most of us are convinced that quantum chromodynamics does explain the way nuclei behave. We say it explains it "in principle," but I am not really sure of what we mean by that.

Still, relying on this intuitive idea that different scientific generalizations explain others, we have a sense of direction in science. There are arrows of scientific explanation, which thread through the space of all scientific generalizations. Having discovered many of these arrows, we can now look at the pattern that has emerged, and we notice a remarkable thing: perhaps the greatest scientific discovery of all. These arrows seem to converge to a common source! Start anywhere in science and, like an unpleasant child, keep asking "Why?" You will eventually get down to the level of the very small.

By the mid-1920s, the arrows of explanation had been traced down to the level of the quantum mechanics of electrons, photons, atomic nuclei, and, standing somewhat off in the corner, the classical theory of gravity. By the 1970s we had reached a deeper level— a quantum field theory of particles called quarks, leptons, and gauge bosons, known as the Standard Model, and, with gravity still somewhat isolated, described by a not very satisfactory quantum field theory of gravitation. The next step, many of us think, is the theory of superstrings, still under development. I myself, although a latecomer to this field, confess my enthusiasm for it. I think it provides our best hope of making the next step beyond the standard model.

Now reductionism, as I've described it in terms of the convergence of arrows of explanation, is not a fact about scientific programs, but a fact about nature. I suppose if I had to give a name for it, I could call it objective reductionism. It is very far from a truism. In particular, these arrows of explanation might have led to many different sources. I think it's important to emphasize that, until very recently, most scientists thought that that was the case. This discovery, that the arrows of explanation point down to a common source, is quite new. (In a comment on an earlier draft of this talk, Ernst Mayr informs me that what I call "objective reductionism" is what he means by "theory reductionism." Maybe so, but I prefer to keep the terms separate, because I wish to emphasize that what I am talking about here is not the future organization of the human scientific enterprise, but an order inherent in nature itself.)

To underscore this point, I'd like to mention a few examples of the contrary view surviving until well into the twentieth century. The first is biological vitalism, the idea that the usual rules of physics and chemistry need to be modified when applied to living organisms. One might have thought that this idea would have been killed off by the rise of organic chemistry and evolutionary biology in the nineteenth century. However, Max Perutz in his talk at the Schrödinger centenary in London in April reminded us that both Niels Bohr and Erwin Schrödinger believed that the laws of physics as understood in the 1920s and 1930s were inadequate for understanding life.[8] Perutz explains that the problem of the orderliness of life that bothered Schrödinger was cleared up by advances in the understanding of enzymatic catalysis. Ernst Mayr was careful in his book to disavow any lingering attachment to vitalism, as follows: "Every biologist is fully aware of the fact that molecular biology has demonstrated decisively that all processes in living organisms can be explained in terms of physics and chemistry." (Mayr, by the way, is using the word "explained" in exactly the same sense as I am here.)

A second example. Lord Kelvin, in a speech to the British Association for the Advancement of Science around 1900, said, "There is nothing new to be discovered in physics now. All that remains is more and more precise measurement." There is a similar remark of Albert Michelson's that is often quoted.[9] These remarks of Kelvin's and Michelson's are usually cited as examples of scientific arrogance and blindness, but I think this is based on a wrong interpretation of what Kelvin and Michelson meant. The reason that Kelvin and Michelson made these remarkable statements is, I would guess, that they had a very narrow idea of what physics was. According to their idea, the subject matter of physics is motion, electricity, magnetism, light, and heat, but not much else. They felt that that kind of physics was coming to an end, and in a sense it really was. Kelvin could not possibly have thought in 1900 that physics had already explained chemical behavior. He didn't think so, but he also didn't think that was a task for physics. He thought that physics and chemistry were sciences on the same level of fundamentalness. We don't think that way today, but it isn't long ago that physicists did think that way.

I said that these arrows of explanation could have led down to a

8. Perutz, M., in *Schrödinger: Centenary Celebration of a Polymath*, ed. Kilmister, C., p. 234 (Cambridge University Press, Cambridge. 1987).

9. Michelson, A. A., *Light Waves and Their Uses* (1903).

number of separate sciences. They also could have gone around in a circle. This is still a possibility. There is an idea that's not quite dead among physicists and cosmologists, the "anthropic principle," according to which there are constants of nature whose value is inexplicable except through the observation that if the constants had values other than what they have the universe would be so different that scientists would not be there to ask their questions. If this version of the anthropic principle were true, there would be a kind of circularity built into nature, and one would then I suppose have to say that there is no one fundamental level—that the arrows of explanation go round in circles. I think most physicists would regard the anthropic principle as a disappointing last resort to fall back on only if we persistently fail to explain the constants of nature and the other properties of nature in a purely microscopic way. We'll just have to see.

Now although what I have called objective reductionism became part of the general scientific understanding only relatively recently (after the development of quantum mechanics in the 1920s), its roots can be traced hack to Newton (who else?). Newton was the first to show the possibility of an understanding of nature that was both comprehensive and quantitative. Others before him, from Thales to Descartes, had tried to make comprehensive statements about nature, but none of them took up the challenge of explaining actual observations quantitatively in a comprehensive physical theory.

I don't know of any place where Newton lays out this reductionist program explicitly. The closest I can come to it is a remark in the preface to the first edition of the *Principia*, written in May 1686. Newton says, "I wish we could derive the rest of the phenomena of nature by the same kind of reasoning from mechanical principles [I suppose he means as in the *Principia*] for I am induced by many reasons to suspect that they may all depend on certain forces." Perhaps the most dramatic example of the opening up by Newton of the possibility of a comprehensive quantitative understanding of nature is in the third book of the *Principia*, where

Newton reasons that the moon is 60 times farther away from the center of the earth than Cambridge is (either Cambridge) and therefore the acceleration of the moon toward the earth should be less than the acceleration of an apple in Cambridge by a factor of 60 squared, or 3,600. With this argument Newton unites celestial mechanics and observations of falling fruits in a way that I think captures for the first time the enormous power of mathematical reasoning to explain not only idealized systems like planets moving in their orbits, but ultimately everything.

A digression. Since I have been talking about Newton, and also talking about the SSC, a prime example of "big science," I can't resist remarking that Newton himself was involved in big science.[10] In 1710, as president of the Royal Society, Newton by royal command was given control of observations at the largest national laboratory for science then in existence in England, the Greenwich Observatory. He was also given the responsibility of overseeing the repair of scientific instruments by the Master of Ordnance, an interesting connection with the military.

There are many gaps, of course, and perhaps there always will be many gaps in what I have called the chains of explanation. The great moments in the history of science are when these gaps are filled in, as for example when Charles Darwin and Alfred Russel Wallace explained how living things, with all their adaptations to their environment, could develop without any continuing external intervention. But there are still gaps.

Also, sometimes it isn't so clear which way the arrows of explanation point. Here's one example, a small one, but one that has bothered me for many years. We know mathematically that as a consequence of Einstein's General Theory of Relativity there should exist polarized gravitational waves, and therefore when quantized, the theory of gravity should have in it particles of mass zero whose spin around their direction of motion equals two (in units of Planck's constant). On the other hand, we also know that

10. Mark, H., *Navigation* 26, 25 (1979).

any particles of mass zero and spin two must behave as described by Einstein's General Theory of Relativity. The question is, which is the explanation of which? Which is more fundamental, general relativity or the existence of particles of mass zero and spin two? I've oscillated in my thinking about this for many years. At the present moment in string theory the fact that the graviton has mass zero and spin two appears as an immediate consequence of the symmetries of the string theory, and the fact that gravity is described by the formalism of non-Euclidean geometry and General Relativity is a somewhat secondary fact, which arises in a way that is still rather mysterious. But I don't know if that is the final answer. I mention this example just to show that although we don't always know which truths are more fundamental, it's still a worthwhile question to ask, because it is a question about the logical order of nature.

I believe that objective reductionism, reductionism as a statement about the convergence of arrows of explanation in nature, is by now ingrained among scientists, not only among physicists but also among biologists like Ernst Mayr. Let me give an example. Here's a quote from the presidential address of Richard Owen to the British Association in 1858.[11] Owen was an anatomist, generally regarded as the foremost of his time, and a great adversary of Darwin. In his address, Owen says, "Perhaps the most important and significant result of palaeontological research has been the establishment of the axiom of the continuous operation of the ordained becoming of living things." I'm not too clear what precisely Owen means by this axiom. But my point is that today no biologist would make such a statement, even if he or she knew what the axiom meant, because no biologist today would be content with an axiom about biological behavior that could not be imagined to have an explanation at a more fundamental level. That more fun-

---

11. *Edinburgh Review* 11, 487–532; see also Hull, D. L., in *Darwin and His Critics* (Harvard University Press, Cambridge, 1973).

damental level would have to be the level of physics and chemistry, and the contingency that the earth is billions of years old. In this sense, we are all reductionists today.

Now, these reflections don't in themselves settle the question of whether the SSC is worth $4.4 billion. In fact, this might be a difficult problem, if we were simply presented with a choice between $4.4 billion spent on the SSC and $4.4 billion spent on other areas of scientific research. However, I don't think that that's likely to be the choice with which we are presented. There is evidence that spending on "big science" tends to increase spending on other science, rather than the reverse. We don't really know with what the SSC will compete for funds. In any case, I haven't tried here to settle the question of whether or not the SSC should be built for $4.4 billion—it is a complicated question, with many side arguments. All I have intended to argue here is that when the various scientists present their credentials for public support, credentials like practical values, spinoff, and so on, there is one special credential of elementary particle physics that should be taken into account and treated with respect, and that is that it deals with nature on a level closer to the source of the arrows of explanation than other areas of physics. But how much do you weigh this? That's a matter of taste and judgment, and I'm not paid to make that final decision. However. I would like to throw into the balance one more point in favor of the SSC.

I have remarked that the arrows of explanation seem to converge to a common source, and in our work on elementary particle physics we think we're approaching that source. There is one clue in today's elementary particle physics that we are not only at the deepest level we can get to right now, but that we are at a level which is in fact in absolute terms quite deep, perhaps close to the final source. And here again I would like to quote from myself, from my own testimony in Congress, because afterward I am going to quote some comments on these remarks, and I want you to know what it is that the comments were about:

There is reason to believe that in elementary particle physics we are learning something about the logical structure of the universe at a very, very deep level. The reason I say this is that as we have been going to higher and higher energies and as we have been studying structures that are smaller and smaller we have found that the laws, the physical principles, that describe what we learn become simpler and simpler. I am not saying that the mathematics gets easier, Lord knows it doesn't. I am not saying that we always find fewer particles in our list of elementary particles. What I am saying is that the rules that we have discovered become increasingly coherent and universal. We are beginning to suspect that this isn't an accident, that it isn't just an accident of the particular problems that we have chosen to study at this moment in the history of physics, but that there is simplicity, a beauty, that we are finding in the rules that govern matter that mirrors something that is built into the logical structure of the universe at a very deep level. I think that this kind of discovery is something that is going on in our present civilization at which future men and women and not just physicists will look back with respect.

After I made these remarks there were remarks by other witnesses, and then there were questions from members of the Committee on Space, Science, and Technology. I am going to quote from the remarks of two of them. The first is Harris W. Fawell, Republican congressman from Illinois. Fawell throughout his questioning had been generally favorable to the SSC. The second is Representative Don Ritter, of Pennsylvania, also a Republican, who had been the congressman most opposed to the SSC throughout the morning. (I suppose you could regard this as a modern dialogue between Salviati and Simplicio.) I quote here from the unedited transcript of the hearings.

Mr Fawell: Thank you very much. I appreciate the testimony of all of you. I think it was excellent. If ever I would want to explain to one and all the reasons why the SSC is needed I am sure I can go to your testimony. It would be very helpful. I wish sometimes we have some one word that could say it all and that is kind of impossible. I guess perhaps Dr. Weinberg you came a little close to it and I'm not sure but I took this down. You said you suspect that it isn't all an accident that there are rules which govern matter and I jotted down, will this make us find God? I'm sure you didn't make that claim, but it certainly will enable us to understand so much more about the universe?

Mr Ritter: Will the gentleman yield on that? [That's something congressmen say to each other.] If the gentleman would yield for a moment I would say . . .

Mr Fawell: I'm not sure I want to.

Mr Ritter: If this machine does that I am going to come round and support it.

While this dialogue was going on I thought of a number of marvelous observations that I could make to score points for the SSC. However, by the time Mr. Ritter reached his final remark I had decided to keep my mouth shut. And that, my friends, is what I learned about the art of congressional testimony.

ACKNOWLEDGMENTS

I was greatly aided in preparing the talk at Cambridge and the article derived from it by conversations with G. Holton, H. Mark, E. Mayr, and E. Mendelsohn. I am also grateful for helpful comments on an earlier written version by J. Krumhansl, E. L. Goldwasser, E. Mayr, M. Perutz, and S. Wojicki.

3

Newton's Dream

Here's another talk given at a 1987 tercentenary celebration of Newton's *Principia*, this one sponsored by Queen's University, the Royal Military College of Canada, and the Royal Society of Canada. Once again I credited Newton with the foundation of the reductionist tradition in modern science, but in this talk I emphasized his commitment to atomism. Of course, modern physics has gone beyond Newton's view of nature as particles moving under the influence of forces acting at a distance. Starting in the nineteenth century, electromagnetic and gravitational forces became attributed to fields; instead of saying that the sun and the earth exert gravitational forces on each other, we say that the sun and earth and other masses create a gravitational field, which exerts gravitational forces on all these bodies. The field interpretation of electromagnetism and gravitation was an essential element in the development in the early twentieth century of Special and General Relativity. We have gone on since then to attribute the strong and weak nuclear forces to fields, and have replaced the particles on which the forces act with fields—electron fields, quark fields, and so on.

Our vision now differs from Newton's in another respect. He believed in the Bible as a source of prophecy, in a way that few scientists do now. Often Newton's commitment to religion is mentioned to rebuke the widespread religious skepticism of today's scientists. I reach a quite different conclusion from Newton's piety. The fact that Newton and Michael Faraday and other scientists of the past were deeply religious shows that religious skepticism is not a prejudice that governed science from the beginning, but a lesson that has been learned through centuries of experience in the study of nature.

———————

This symposium has been organized to celebrate a great book published three hundred years ago, the *Principia* of Isaac Newton. In that book Newton outlined a new theory of motion and a new theory of gravity, and succeeded thereby in explaining not only the apparent motions of bodies in the solar system, but terrestrial phenomena like tides and falling fruits as well. In other work Newton developed the mathematics of the calculus.[1] Newton also performed fundamental experiments in the theory of optics and wrote books about biblical chronology. Yet with all these accomplishments Newton can be said to have contributed to our species one great thing that transcends all his other specific scientific achievements. The title of my talk expresses it: "Newton's Dream."

Newton's dream, as I see it, is to understand all of nature, in the way that he was able to understand the solar system, through principles of physics that could be expressed mathematically. That would lead through the operation of mathematical reasoning to predictions that should in principle be capable of accounting for everything. I do not know of an appropriate place in the corpus of Newton's writings to look for a statement of this program. Newton scholars confirm to me that Newton had this aim, but the closest to an explicit statement of it I have found is in the preface to the first edition of the *Principia*, written 301 years ago, in 1686: "I wish we could derive the rest of the phenomena of nature [that is, the phenomena that are not covered in the *Principia*] by the same kind of reasoning as for mechanical principles. For I am induced

———————

1. He was apparently not entirely comfortable with calculus, and played it down in the *Principia*, but it is the method by which today we carry out Newton's calculations. It is very hard for a modern physicist to read the *Principia* because its style is so geometric.

by many reasons to suspect that they may all depend on certain forces." He wanted to go on beyond the *Principia* and explain everything.

For many years after Newton this ambition took the form of atomism. Clearly, it was a major challenge to explain the behavior of ordinary materials, not in the solar system, where the forces are simple—just the long-range force of gravity, but also here on Earth where things are messy, where axletrees get stuck in the mud, soup boils on the stove, and an apple left on a windowsill rots. He wanted to explain all of these phenomena by forces like gravity acting on the elementary particles out of which matter is composed.

There is a nice statement of this at the end of his second great book, *The Optics,* first published in 1704.[2] At the end of *The Optics* Newton begins to philosophize about the future of science: "All these things being considered, it seems probable to me that God in the beginning formed matter as solid, massy, hard, impenetrable, moveable particles of such sizes and figures and with such other properties and in such proportions of space as most conduced to the end for which he formed them"; and he goes on later: "It seems to me further that these particles have not only *Vis Inertiae* [which I gather is translated as kinetic energy] accompanied with such passive laws of motion as naturally result from that force, but also that they are moved by certain active principles such as that of gravity and that which causes fermentation and the cohesion of bodies. These principles I consider not as occult qualities but as general laws of nature."

His picture was that the universe consists of particles, the particles are elementary, they cannot be subdivided, they are eternal, but they act on each other through forces: these forces include the force of gravity, but other forces also. Newton was not so foolish as to think that the processes of life or the flow of fluids or the ordinary properties of everyday matter could all be explained in

2. I quote here from a later edition.

terms of gravity. He knew there must be other forces. But he hoped that these other forces could be discovered and that then the behavior of matter could be understood through the action of these forces on the elementary particles of matter.

The tradition of atomism, of course, was very old. It predates Newton by several millennia. It goes back to the town of Abdera in ancient Greece, where Democritus and Leucippus had the same conception. But although many before Newton had the idea of explaining everything in simple terms, from Democritus and Leucippus to Descartes, Newton was the first to show how it would work—to show with the example of the solar system how one could explain the behavior of bodies mathematically and make predictions that would then agree with experiment. Others, as I said, had these grand hopes. René Descartes, who slightly preceded Newton and from whom Newton learned much, also had a broad, general view of nature. But Descartes somehow or other fell short. He never quite grappled with the task of applying quantitatively his principles to the prediction of phenomena. The Newtonian approach—the Newtonian success—had no predecessors, and it left physicists with the challenge of carrying it further.

Atomism continued to be at the center of this program for many scientists. In the early nineteenth century great steps were taken in applying the atomic idea to chemistry. Generally speaking, chemists in the nineteenth century became quite comfortable with atoms: they measured atomic weights and knew that there were molecules of water, for instance, which consisted of three atoms, two of hydrogen and one of oxygen. However, atoms seemed hopelessly remote from direct observation. I would imagine that not only Democritus and Leucippus, but also even John Dalton and Amadeo Avogadro, felt that atoms were as far removed from their experimental direct observation as we now think superstrings are removed from direct study. And, in fact, by the end of the nineteenth century there was to some degree a reaction against atomism, and against Newton's dream. Several German and Austrian physicists, led by Ernst Mach, followed a positivistic line, according to which physicists are supposed simply to make measurements. Theory was admitted to be convenient as a method of organizing the measurements, but physicists were not supposed to inquire more deeply into things; in particular they were not supposed to ask about atoms, which are not directly measurable.

Another challenge to the atom came from the development of an alternative, the field. Again starting in England with the work of Faraday and continuing with the great synthesis of James Clerk Maxwell, it became possible to think that perhaps the fundamental ingredients of nature are not atoms at all but fields, extended regions within which there is energy and momentum that perhaps form tight knots that we observe as atoms. Maxwell's formulation of the field theory of electricity and magnetism is in its way as much of an achievement as Newton's mathematical theory of the motion of bodies under the influence of gravity, except of course that Maxwell followed in Newton's footsteps.

When the electron was discovered in 1897, the same experiments were being done in England and in Germany. In England there was a predisposition, because of the tradition of Newton, William Prout, and Dalton, to believe in elementary particles, and when these experiments were performed at the Cavendish Laboratory, by J. J. Thomson in 1897, they were immediately heralded as the discovery of the electron. The same experiments were being done in Berlin by Walter Kaufmann, who simply reported that he had observed a cathode ray to bend in a certain way under the influence of electric and magnetic fields.

The atomistic idea had its first triumph with the discovery of the electron, but the field idea was still very strong, and immediately field theorists went to work trying to make models of the electron. The early twentieth century saw a number of extremely elaborate models by Jules-Henri Poincaré, Max Abraham, and others in which the electron was supposed to be just a little bundle of field energy. It was the particular genius of Einstein in 1905 to realize that the time was not yet right for a model of the electron, that in fact one should try to understand the motion of electrons and the behavior of light in terms of symmetry principles, particularly his famous principle of relativity, and put off for the future the question of the nature of the electron.

I have been discussing Newton's followers only in the context of physics, although in the nineteenth century great steps were also made to realize Newton's dream outside of physics. Perhaps even more important for the intellectual evolution of Western civilization were the developments in biology, above all the realization by Darwin and Alfred Wallace that life, with all its apparent purpose and its apparent fashioning for use, can develop through a more or less random series of breedings and eatings. This fact, together with the discovery that organic chemicals can be synthesized from inorganic chemicals, led to the realization that life is not something apart from the same general world of phenomena that, following Newton's dream, may some day be explained. As Newton said: "They may all depend on certain forces."

These ideas came together and began to make sense with the advent of quantum mechanics in the 1920s. With that, we understood for the first time what atoms are and how the forces that act within atoms, specifically the electromagnetic force, produce all the rich variety of chemical behavior which, over the course of billions of years, produced the phenomenon of life. And although Newton started the dream, and many great contributions were made to it in the ensuing centuries, the beginning of the realization of Newton's dream is the advent of quantum mechanics and its explanation of the nature of ordinary matter.

By the mid-1920s one could say that in some sense all natural phenomena could be explained, at least in principle, in terms of the particles that make up ordinary matter. These particles were at that time conceived to be electrons, which are the particles in the outer parts of atoms, discovered by Thomson (and if he had wanted to claim credit for it, Kaufmann); photons, the particles that in large numbers make up a ray of light; and atomic nuclei, which were then mysterious, positively charged masses at the centers of atoms that hold the electrons in their orbits by the force of electrical attraction, and that contain by far the greatest part of the mass of ordinary atoms. And then one other ingredient in the universe, not understood at all, seeming to have nothing to do with phenomena at the atomic scale: the force of gravity.

The physicists of the 1920s had what seemed a fairly simple world. They didn't understand anything about nuclei. They knew about protons, which are the nuclei of the simplest atom, the atom of hydrogen, and which are particles with just one positive unit of electric charge. They knew that other nuclei could not just consist of protons, because they had a smaller ratio of charge to mass than protons. And they knew that every once in a while nuclei spit out electrons. It seemed obvious to them that nuclei consist of protons and electrons. There were terrible problems, theoretical problems, in trying to make a picture of a nucleus consisting of protons and electrons holding together, but it was hoped that somehow all would be understood in the future. I suppose the physicists of the 1920s would have said the universe consists of electrons, protons, electromagnetism, and also gravitation.

There is a famous remark made just after Paul Dirac formulated his relativistic theory of the electron in 1928. Someone, whose name I cannot recall, said: "In two more years, we will have the proton [meaning we will have the relativistic quantum theory of the proton] and then we will know everything." Well, that didn't happen.

The 1930s saw a period of intense study of atomic nuclei. It was discovered that nuclei consist not of protons and electrons but of protons and other particles called neutrons. After World War II, when a new generation of accelerators came on line, it began to be possible to produce entirely new species of particles, particles that seemed like siblings of the electron or the proton or the neutron, but were heavier. And because they were heavier (so that there was energy available) they could easily decay into the lighter particles and did so. They lived very short lives, and so had to be artificially produced in laboratories at Berkeley and Brookhaven. The years after the mid-1950s saw an attempt to come to grips with the expanding population of species of particles, and by the mid-1970s it had fallen into place with the formulation of what is now called the "Standard Model."[3] The formulation of the Standard Model of weak, electromagnetic, and strong interactions was finished by the mid-1970s, and by then there were enough experiments to show that one could actually rely on it. In the Standard Model the list of fundamental ingredients of the universe still includes the electron, but also its siblings, particles which are quite like the electron and which can turn into electrons under the influence of the weak force, particles called neutrinos, and muons, and tauons. The proton and the neutron no longer are seen to be basic. They are seen as mere bound states, something like an atom, or a molecule, or a blackboard eraser, made up of more elementary particles. The elementary particles of which the protons and the neutrons are supposed to be composed are known as quarks. A number of different families of quarks appeared that had to be named. The first two were called Up and Down, because one had a positive and the other a negative charge. The quarks that followed were called Strange, because they are present in particles whose discovery had come as a surprise, and then Charm for no clear reason, and then Top and Bottom. (I co-authored a very speculative paper during this period in which there happened to appear seven types of quarks; we decided to call them Gluttony, Envy, Sloth, and so on.)

The electron and other similar particles collectively are called leptons. There is apparently a parallelism between quarks and leptons. There are exactly the same number of flavors of quarks and leptons and we think we know why that is true. The particles transmit the forces. We have known about the photon ever since

3. "Standard Model" is a useful term. I have used it in a book on gravitation and cosmology to describe the Big Bang theory of the universe, and I understand that biologists also have a standard model. This term signifies a theory that we take as our working hypothesis, a hypothesis about which it is respectable to write scientific papers, but which by no means are we absolutely committed to believe.

Einstein proposed it in 1905, but we now know the photon has much heavier siblings that transmit the weak nuclear force. The W particle and the Z particle, as they are called, were discovered only in 1983 at CERN, the elementary physics laboratory in Geneva. There is another similar class of particles that carry the strong nuclear force. They are called gluons because they glue the quarks together inside the proton and neutron and, as a by-product, also glue the proton and neutron together inside the atomic nucleus.

The photon is a sibling of the W and the Z, but they have vastly different masses; even though there is a family relation among them they seem very different. We now understand this in terms of a phenomenon called spontaneous symmetry breaking. This refers to the fact that a set of mathematical equations can have a high degree of symmetry (that is, can retain the same form when the different variables are transformed into each other), and yet the solutions of those equations may not share that symmetry. On the level of the underlying equations that govern the particles, the photon, the W, and the Z all appear symmetrical. There are transformations that change a photon-Z mixture into a W and leave the form of the equations unchanged. Yet the solutions of the equations, which are the particles themselves, don't exhibit that symmetry. They are very different. Once one sees through the breaking of the symmetry, one can easily understand the particles.

We do not yet know, however, the mechanism by which the symmetries are broken. A similar phenomenon occurs in superconductivity (in fact, that is where it was discovered first). Superconductivity is on everyone's mind these days, but elementary particle physicists in the late 1950s had already learned much about it from their brethren in solid state physics. In the theory of superconductivity, the symmetry that is broken is ordinary electromagnetic gauge invariance, or to put it more simply, conservation of electric charge. This symmetry is broken by forces that are transmitted by sound waves (or, in a quantum sense, phonons) between the electrons in the superconductor. The question is, what plays the role of the sound waves in elementary particle physics?

What are the forces that induce the instability, which produces the breakdown of the symmetry? We don't know. There are some simple pictures of these forces. The simplest involve certain particles that have come to be known as Higgs particles because they appeared in illustrative mathematical models invented by Peter Higgs of Edinburgh and others. This is the part of the Standard Model that we are hoping will be clarified by experiments at the great new accelerator that we hope will be built in the next six or seven years, the Superconducting Super Collider (or SSC).

It is clear that the Standard Model is not the final realization of Newton's dream. Even when we get straight about these Higgs particles there will still be too many arbitrary features in the Standard Model. Even if there were no particles waiting to be discovered besides those which we know are theoretically necessary—specifically one more quark, the top quark, and the Higgs boson—then even so, the Standard Model will have in it eighteen numerical quantities like the charge of the electron, the mass of the electron, the masses of the quarks, and so on. We learned that they have the values they have from experiment, but we don't know why nature chooses those values. Any theory that has eighteen free parameters is too arbitrary to be satisfactory.

There has to be something else. The Standard Model leaves out gravity. Gravity is still doing its job in the solar system, the way it was in the 1920s, having no observable effect at the level of atoms and molecules, and being absolutely inaccessible to experimental study except on macroscopic scales, where we have enormous numbers of particles adding up their gravitational fields. We need a theory that goes beyond the Standard Model because we want to explain those eighteen parameters. More than that, we want to explain the Standard Model itself. We want to explain why there have to be all these flavors of quarks and leptons. And of course we want to bring gravity into the picture.

Many theorists feel that the next step will take the form of superstring theories, in which the basic ingredients of the universe are seen not to be particles, not to be fields, but instead to be something like rubber bands, little strings that go zipping around, vibrating in many normal modes. The normal modes of vibration are what we see as different species of particles, but they are all one kind of string. This kind of theory has a long way to go before it is able to explain things like the values of the eighteen parameters of the Standard Model.

I have taken as landmarks in this discussion the 1920s synthesis in terms of the quantum mechanics of electrons and protons and the 1970s synthesis of the Standard Model, so by an arithmetical progression, we really shouldn't expect any new breakthroughs until 2020. Edward Witten has said that the reason we are having so much trouble doing superstring physics is that it is twenty-first-century physics, which we accidentally stumbled on thirty years too early.

The shape of the final realization of Newton's dream is very far from clear. We believe, though, that it will have to include quantum mechanics. Since the development of quantum mechanics in the 1920s there has been not the slightest suggestion that any physical phenomena require any correction to quantum mechanics. And it is always the quantum mechanics of 1925 to 1926, the quantum mechanics we learn in our first course. The quantum mechanics worked out by Erwin Schrödinger and Werner Heisenberg and Max Born and Pascual Jordan and Wolfgang Pauli in 1925 to 1926 is the same quantum mechanics we apply to quantum fields in the Standard Model, and it is the same we apply to superstrings today. Quantum mechanics seems to be a permanent part of our physical understanding.

Lately I have been trying to understand why that is, by asking what kind of generalization of quantum mechanics could possibly be logically consistent. Is there any way, for example, of taking the linear equations of quantum mechanics and introducing nonlinearities? I can tell you it is very hard. It is very hard to think of any way of tampering with the rules of quantum mechanics without having logic fly out the window, for example, without introducing negative probabilities or probabilities that don't add up to one. So I find it difficult to imagine that the future synthesis of everything will not be in the language of quantum mechanics.

Quantum mechanics is a grammar, in terms of which all physics must be expressed, but it does not itself tell us anything. That's one of the reasons it is so hard to test, because by itself it says nothing. The other ingredient that seems to be needed to add to quantum mechanics to complete a picture of the universe is the symmetries. Now that may surprise you because in many areas of science symmetries are somewhat incidental. Biologists, for instance, know that a human being has roughly a symmetry that the mathematician calls Z_2, the interchange of right and left, but that is certainly not the most interesting thing about human beings. In particle physics, on the other hand, it seems that symmetries are the most interesting things about elementary particles. Think of what it takes to describe an elementary particle. One describes an elementary particle by giving its momentum and its energy and its charge and a few other things. Every one of these is simply a number that describes how the wave function (the mathematical object representing a physical state in quantum mechanics) changes when symmetry transformations are performed. Aside from their symmetry transformation properties, all particles are the same. There may be nothing to nature but quantum mechanics, which is the stage for physical phenomena, and the symmetry principles, which are the actors.

But that's looking far ahead. We do not know that there aren't entirely new ideas, which will have to be invented. The progress in superstring theory has not been as exciting in 1987 as it was in 1984 or 1985. And it is beginning to look as if some very different new ideas are needed to make further progress there.

Now that I have given you a complete history of science in the last three hundred years, I would like, if I may, to try to draw some lessons from this experience. Those of us whose lives are governed by our own little versions of Newton's dream are trying to accomplish something very special. Our work, although we are delighted if it has some utility, is not particularly directed toward utility. Nor have we chosen the problems we are working on because they are fun or mathematically interesting. Sometimes we are accused of that, and in fact, sometimes, as a kind of self-protection and to avoid the accusation of taking ourselves too seriously, we claim that we do the work we do just for fun. But that is not all there is to it. We, meaning the community of elementary particle physicists and those in the related disciplines of cosmology and astrophysics, have a historical goal in mind. The goal is the formulation of a few simple principles that explain why everything is the way it is. This is Newton's dream, and it is our dream.

This sense of historical direction makes our behavior somewhat different from that of scientists in general. For one thing, the importance of phenomena in everyday life is, for us, a very bad guide to their importance in the final answer. The fact that the electron is ubiquitous in ordinary matter is simply because the electron is lighter than the muon. The muon is 200 times heavier than the electron, so there is lots of energy available for the muon to decay into the electron and a couple of neutrinos, which it does with a lifetime of a microsecond or so. We do not know about muons in our everyday life. But as far as we know, muons play just as fundamental a role (which may or may not be very fundamental) as electrons in the ultimate scheme of things. The fact that electrons were discovered first and are far more common in matter throughout the universe is of secondary importance.

This point comes up because often, although muons are present in cosmic rays, by and large these particles are particles that we have to produce artificially in the laboratory. One might argue that we particle physicists are like butterfly collectors, studying butterflies that we have created ourselves in our laboratory and which do not exist in the real world. But we are not very interested in our butterflies. We are not particularly interested in our electrons or our muons. We are interested in the final principles that we hope we will learn about by studying these particles. So the first lesson is that the ordinary world is not a very good guide to what is important.

I suppose this lesson could have been learned before. Copernicus, you remember, thought that the orbits of planets had to be perfect circles because something as important as a planet would have to move on a perfect curve, such as a circle. Today we don't look at planets as being very special. Planets are just bodies that happen to have formed in the history of the solar system and whose orbits have various eccentricities, but there is no particular reason why their orbits have to be circles.

This brings me to the second lesson. It is that if we are talking about very fundamental phenomena, then ideas of beauty are important in a way that they wouldn't be if we were talking about mere accidents. Planetary orbits don't have to be beautiful curves like circles because planets are not very important on any fundamental level. On the other hand, when we formulate the equations of quantum field theories or string theories we demand a great deal of mathematical elegance, because we believe that the mathematical elegance that must exist at the root of things in nature has to be mirrored at the level where we are working. If the particles and fields we were working on were mere accidents that happened to be important to human beings, but were not themselves special, then the use of beauty as a criterion in formulating our theories would not be so fruitful.

Finally, the kind of beauty for which we look is special. Beauty, of course, is a general and broad and vague word. We find many things beautiful. The human face is beautiful, a grand opera is beautiful, a piano sonata is beautiful. The kind of beauty we are looking for is more like the beauty of a piano sonata than that of a grand opera, in the specific sense that the theories we find beautiful are theories which give us a sense that nothing could be changed. Just as, listening to a piano sonata, we feel that one note must follow from the preceding note—and it could not have been any other note—in the theories we are trying to formulate, we are looking for a sense of uniqueness, for a sense that when we understand the final answer, we will see that it could not have been any other way. My colleague John Wheeler has formulated this as the prediction that when we finally learn the ultimate laws of nature we will wonder why they were not obvious from the beginning.

That may very well be true. If it is, I suspect it will be because by the time we learn the ultimate laws of nature, we will have been changed so much by the learning process that it will become difficult to imagine that the truth could be anything else but superstring theory—or whatever it turns out to be.

So far, I have been speaking in a style that is sometimes called physics imperialism. That is, the physicist provides a set of laws of nature that explain everything else and all the other sciences appear to be offshoots of physics. I want, at least in part, to disavow this. I do believe there is a sense in which everything is explained by the laws of nature and the laws of nature are what physicists are trying to discover. But the explanation is an explanation in principle of a sort that doesn't in any way threaten the autonomy of the other sciences. We see this even within physics itself. The study of statistical mechanics, the behavior of large numbers of particles, and its applications in studying matter in general, like condensed matter, crystals, and liquids, is a separate science because when you deal with very large numbers of particles, new phenomena emerge. To take an example I have used elsewhere, even if you tried the reductionist approach and plotted out the motion of each molecule in a glass of water using equations of molecular physics to follow how each molecule went, nowhere in the mountain of computer tape you produced would you find the things that interested you about the water, things like turbulence, or temperature, or entropy. Each science deals with nature on its own terms because each science finds something else in nature that is interesting. Nevertheless, there is a sense that the principles of statistical mechanics are what they are because of the properties of the particles out of which bodies are composed. Statistical mechanics does not have principles that stand alone and cannot be deduced from a deeper level.

As it happens, that deeper level has usually been a more microscopic level. If we ask any question about nature—why the sky is blue or the grass is green—and keep asking why, why, why, we will get a series of answers that generally takes us down to the level of the very small.

There is a sense in which the kind of thing that elementary particle physicists study is especially fundamental, but in no way does it threaten the separate existence or the special importance of other sciences. Because we physicists think we are moving toward the final answer, the work we are doing is not necessarily more worthy of support than the work of other scientists. This has come up in Britain with the debate over further participation in CERN, and recently in the United States in the debate about whether or not to spend $4.4 billion for the SSC accelerator. It is argued by the opponents of CERN in Britain and the opponents of the SSC in the United States that elementary particle physics is no more fundamental than other areas of science; it is less likely to yield results of direct practical importance, and therefore it should not take such a large share of the public funds of these countries. These are very difficult questions, and I would have a hard time myself deciding which areas of science deserve what part of a research budget. It may not be an essential issue, however, because there is evidence that spending on large research projects like the SSC actually helps to buoy up spending on research in general. But I don't want to get into that argument because it's a matter in which I don't have any expertise.

I want to make only one point about the funding of elementary particle physics, but that point I want to make very strongly. Although all of the sciences have credentials to justify their support, credentials that may take the form of practical utility, or impact on neighboring sciences, or intellectual challenge, there is one particular credential that elementary particle physics has that is not necessarily more important than the others but that *is* worthy of respect. And that is that we are trying to get to the roots of the chains of explanation, that we are trying to grapple with nature at the most fundamental level that it is given to human beings to address. In other words, that it is we who are trying to realize Newton's dream.

4

Confronting O'Brien

When I was a boy, before the days of television, there was a popular radio program called "Information Please." A few literate and articulate gentlemen answered questions on scholarly topics put by the moderator, Clifton Fadiman. (I never knew the answers.) It was quite a surprise when, in 1988, I received a letter from Mr. Fadiman, asking me to contribute to *Living Philosophies*. As he described it, this would be a book containing statements of personal philosophy by an interesting assortment of men and women. Some were friends, like Daniel Boorstin, Freeman Dyson, Stephen Gould, and E. O. Wilson, and there were others whom I did not know but had admired, including Abba Eban, Jane Goodall, Walker Percy, and Stephen Spender. I of course accepted; to me, Fadiman's letter was a message from out there in the radioland of my boyhood. Besides, no one had ever asked me before for my personal philosophy.

My essay turned out to be mostly concerned with epistemology: how should we decide what to believe? Toward the close of the essay I had a little to say about what I do believe, or really, as I see in rereading it, about what I do not believe. I do not believe in a cosmic plan in which human beings have any special place, or in any system of values other than the ones we make up for ourselves. I ended with a description of our world as a stage, onto which we have stumbled with no script to follow. This was not the first time that I had borrowed Shakespeare's metaphor of the world as theater; my 1977 book *The First Three Minutes* ends with "The effort to understand the universe is one of the few things that lifts human life a little above the level of farce, and gives it some of the grace of tragedy." But the tragedy is not in the script; the tragedy is that there is no script.

For many years I have been a cheerful philistine in philosophical matters. Still (like everyone else) I worry about the Big Questions from time to time, and the invitation to contribute to this book provides an occasion to pull my thoughts into order. I will offer some remarks about one Very Big Question: how should we decide what we ought to believe? Then I will say somewhat less about what I do believe.

When it comes to issues of fact rather than of value, I take it as a point of honor, as a moral rather than a logical necessity, to judge matters by the methods of science, or by their commonsense analogs in everyday life. This statement may seem both stuffy and banal, but I think it has serious implications for the way one lives, and it raises enough problems to require some immediate qualifications.

First, the qualifications. I know enough about science to know that there is no such thing as a clear and universal "scientific method." All attempts to formulate one since the time of Francis Bacon have failed to capture the way that science and scientists actually work. Still, under the general heading of scientific method, we can understand that there is meant a commitment to reason, often though not necessarily crystalized as mathematics, and a deference to observation and experiment. Above all, it includes a respect for reality as something outside ourselves, that we explore but do not create. The realism of the working scientist is greatly strengthened by the experience (one that has often been my privilege to enjoy) of finding one's theoretical preconceptions overturned by experimental data.

Another qualification: I have made a point of limiting my remarks to matters of fact rather than of value, because it seems to me that there is an unbridgeable gulf between the "is" and the "ought." Indeed, I am suspicious of all efforts from Aristotle on to be systematic about ethics or aesthetics, and even more skeptical about attempts actually to prove anything about such matters. However, I admit that I don't know how to spell out in advance precisely what I mean by the difference between facts and values. It doesn't seem to matter—I can't think of a case where the distinction is not reasonably clear.

Finally, though I write here about how we should decide what we ought to believe, it is not so clear to what extent we *can* decide what to believe. For instance, it is irresistible to believe in our own existence, and most of us cannot help believing in the existence of the world around us, even under the urgings of the most skeptical philosophers. Nevertheless, it seems to me that we can exert some control over our standards of belief, and that we ought to try to follow something like the methods of science. But it is not so easy, and perhaps not so common.

Throughout history, people seem to have been willing to accept authority in judging matters of fact, either the authority of the living, organized in governments and sects, or the authority of the dead, expressed in tradition and sacred writings. I don't mean that they have bowed to authority just in what they say they believe (which could be set down to a reasonable prudence), but also in what they *do* believe. George Orwell's novel *1984* gives a brilliant description of how this can work. His hero, Winston Smith, had written that "freedom is the freedom to say that two plus two makes four." The inquisitor, O'Brien, accepts this as a challenge. Under torture, Smith is easily persuaded to say that two plus two makes five, but that is not what O'Brien is after. As the pain increases, Smith becomes so eager to stop his torture that he manages to convince himself for a moment that two plus two could be five. O'Brien is satisfied, and the torture ends for a while. Authority does not usually operate with such efficiency, but even when it threatens us with smaller pains, over a lifetime these can have impressive effects on what we believe.

Today the hold of authority on our beliefs seems to be slipping, at least in the more developed parts of the world. However, we needn't rely on outside authority as a source of pain. We all have to face the prospects of our own deaths, and the deaths of those we love. There is pain enough for us each to be our own O'Briens.

I can recognize the existence of these sources of belief, but I don't really understand them. When the second triumvirate declared officially in 42 B.C.E. that Julius Caesar was a god, I can well understand that prudent citizens would pay their respects to his statue, but how could they believe that he really had become divine? And if they did not believe it, what was the point of it all? Today, with the decline of authority, one often hears people explain that they believe in this or that because it makes them better or happier. It's easy to see how one's beliefs could arise from a desire to be good or happy, but if one *knows* that this is the source of one's beliefs, how can one continue to believe? I suspect that belief is not the simple matter discussed by epistemologists, but a complicated emotion like love, and that many find it satisfactory to believe and not believe the same thing at the same time. But I think one ought at least to try to resist this temptation.

I would not try to justify the scientific method where it applies by appeals to reason or experience—all such arguments are hopelessly circular. All I can offer are my own feelings, that there is a moral imperative when it comes to deciding matters of fact to decide them in this way; that there is something ignoble in doing otherwise.

Rereading the last few paragraphs, I am bothered by their tone of self-satisfaction. I do not know to what extent my beliefs are under my control. To the extent that they are, why should I follow the cold rules of science, rather than believe in what would help to make me good or happy? And if I cannot adjust my beliefs to suit myself, why should that be a source of pride? After all, I take no pride in the fact that I can't consciously control my heartbeat or my knee-kick reflex. Still, there it is—I am proud that I will not or can not believe in what I do not believe.

So, after this long preamble, what *do* I believe? All my experience as a physicist leads me to believe that there is order in the universe. Here and there amid the general messiness of ordinary life we see order appearing spontaneously, as in a crystal or a rainbow. There is also the grand order of the expansion of the whole universe, and, more intricate, the adaptive orderliness of living things, evolved over hundreds of millions of years of natural selection. And underlying all these is an order in the logical structure of the universe. We can ask why crystals or rainbows or DNA are the way they are, and receive an answer in terms of the physics of atoms and radiation, and if we ask why these physical principles are the way they are, we get an answer in terms of a quantum field theory of particles called quarks, leptons, and gauge bosons. As we proceed down this chain of explanation, the principles seem to become simpler and more unified. We explain more and more with less and less input. I think, though I am not sure, that at the end of the process we will find a few general principles, laws of nature, of great simplicity and beauty.[1] At any rate, I have devoted my own life to work in elementary particle physics and cosmology because I think that, at this moment in history, this is the best way to make progress toward these final laws of nature.

It would be wonderful if, in uncovering the principles underlying the universe, we had discovered ourselves; if we had found that a grand cosmic drama was going on, in which human beings play a starring role. We haven't. Nothing that scientists have discovered suggests to me that human beings have any special place in the laws of physics or in the initial conditions of the universe.[2]

1. I have written at greater length on this point in the second essay in this collection.

2. There is a possible exception. Physicists sometimes speculate about an "anthropic principle," according to which the laws of nature or the constants of nature are what they are at least in part because otherwise there would be no intelligent beings to study them. For a comprehensive account, see J. D. Barrow and F. J. Tipler, *The Anthropic Cosmological Principle* (Oxford: Clarendon Press, 1986). This point of view has been strongly urged by John Archibald Wheeler, and I have toyed with it myself, in *Physical Review Letters* 59 (1987): 2607. In my view, if the anthropic principle has any validity at all, it is only as a means of judging in which part of a mega-universe, an ensemble of what one would ordinarily call universes, we might find ourselves. But it still leaves us with the prospect of an infinitely larger number of other universes in this ensemble that are hopelessly hostile to life, intelligent or not.

If this is true, then we are not going to be able to look to science for any help in deciding what we are to value. At the end of my book *The First Three Minutes*, I allowed myself to remark that "the more the universe seems comprehensible, the more it also seems pointless." This one sentence got me into more trouble with readers than anything else I've ever written, but all I meant was that if we search in the discoveries of science for some point to our lives, we will not find it. This does not mean that we can't find things that give point to our lives. If science can't provide us with values, neither can it invalidate them.

My own values form something of a disorganized hodgepodge. High on the list are loving one another, enjoying beauty, both natural and man-made, and learning about the universe scientifically. I don't see much point in my trying to be systematic about such things, because even if I did organize my values into a coherent system, like utilitarianism or John Rawls' distributive justice, and then found that some imperative of the system contradicted my intuitive value judgments (as is in fact the case for these two examples), I would then just scrap my system. This is not to say that moral systems can never be worthwhile. There are some value judgments that we all want to see imposed on everyone, such as the prohibitions against murder or slavery, and if one is in a position to influence others it might be helpful to organize one's values into a system that has some persuasive power. If it can be rooted in tradition so much the better. But however systematic our values may be, we have to live with the fact that at bottom, in our value postulates, we are not responding to some cosmic imperative, but rather inventing values for ourselves as we go along.

It seems to me that we are in the position of a company of players who have by chance found their way into a great theater.[3] Outside, the city streets are dark and lifeless, but in the theater the lights are on, the air is warm, and the walls are wonderfully decorated. However, no scripts are found, so the players begin to im-

3. This metaphor is of course not new. For a beautiful early version, written long before *Macbeth*, and used for quite a different purpose, see chapter 13 of book 2 of Bede's *Ecclesiastical History of the English People*.

provise—a little psychological drama, a little poetry, whatever comes to mind. Some even set themselves to explain the stage machinery. The players do not forget that they are just amusing themselves, and that they will have to return to the darkness outside the theater, but while on the stage they do their best to give a good performance. I suppose that this is a rather melancholy view of human life, but melancholy is one of the distinctive creations of our species, and not without its own consolations.

5

The Heritage of Galileo

The University of Padua was founded in 1222, and in the Renaissance became the most distinguished university in Europe. In 1592 the university offered a professorship to Galilei Galileo. Galileo was, as we would now say, an upwardly mobile young professor; he quit the university in his home town of Pisa and moved to Padua. Four hundred years later the University of Padua decided to celebrate the anniversary of Galileo's inaugural lecture in December 1992. There was a conference on December 6 on Galileo and his heritage, and an honorary degree ceremony on the seventh. The essay below is essentially the talk I gave at the conference, except for the last four paragraphs, which are taken from my remarks at the ceremony the next day. My lack of enthusiasm for the Vatican's reconciliation with Galileo seemed to make the Italian physicists in the audience happy. I was sorry not to get any reaction from the bishops sitting in the front row; I guess they had heard it all before.

Earlier this week I was attending a seminar on elementary particle physics in Austin, Texas, and found my thoughts drifting to the talk that I was to give here today in Padua. Suddenly I found myself wondering what Galileo would think if he returned to life in the middle of the physics seminar I was attending. He might be depressed by the drabness of the seminar room; he had lived his life in Florence and Pisa as well as Padua, and even under house arrest his surroundings were probably more pleasing than a typical seminar room in any modern university.

On the other hand, Galileo probably would have been very impressed by the overhead projector. He would doubtless have recognized it as a distant descendant of the apparatus he had used in his own study of sunspots, and by the end of the seminar he would probably have figured out a way of improving the projector. He would not have understood the speaker; I don't believe that Galileo understood English (a language of no great international importance in Galileo's time), and of course he did not know anything about elementary particles. Yet I imagine that if one of our graduate students from Italy had translated the proceedings in the seminar and explained the ideas of elementary particle physics to Galileo, he would not have found them entirely strange. After all, Galileo had invented the peculiar blend of theory and experiment that forms the style of our modern science.

Galileo was the first to regard matters of dynamics, like the law of motion of falling bodies, as open questions that must be settled not by pure thought but by quantitative measurements. And though he realized that these measurements could never be exact, at the same time he recognized that the results of these measurements could be integrated into precise mathematical laws. Great as Galileo's discoveries were, it is the invention of this scientific style rather than any specific discovery that constitutes Galileo's greatest legacy to today's science. In the history of science only Newton can be compared to Galileo. Galileo did not have Newton's vision of a comprehensive theory of everything, but Newton reaped the grain that had been planted by Galileo.

Among Galileo's specific discoveries, the law of falling bodies has for physicists a special significance. Galileo was not the first to study the motion of falling bodies quantitatively. Although in Aristotle's *Physics* and *On the Heavens* the problem was addressed in purely qualitative terms, the first three kings of Hellenistic Alexandria subsidized quantitative research on the flight of projectiles. But this was for military purposes, for which one is only interested in the shape and range of a projectile's trajectory. Galileo was the first to bring the functional relation between the projectile's position and *time* into the reach of quantitative study. Indeed, Galileo's discovery that the distance that a projectile falls is proportional to the square of the time appears today as a mere detail, compared with the much greater discovery that there *is* a general relation between distance and time that can be studied quantitatively. Galileo also gave us the pendulum as a means of telling uniform intervals of time, and he found in the revolutions of Jupiter's moons a heavenly clock that could be used to determine longitude. The concern of physicists with the quantitative study of changes in time begins with Galileo, and so it is *time* on which I shall concentrate in this lecture.

For three centuries the nature of time was conceived to be pretty much the same as supposed by Galileo—an inexorable succession of instants that make an absolute separation between past and future. The history of physics in this century has been largely a matter of coming to terms with necessary changes in our view of time. As everyone knows, in the first two decades of the century Einstein's Special and General Theories of Relativity showed that the flow of time could be affected by the motion of an observer or by a gravitational field. In contrast, time appeared in the quantum mechanics of the 1920s in much the same way as it did to Galileo; the wave function like a rolling ball evolves as a function of an objective absolute time.

Matters became far more interesting when quantum mechanics was combined with relativity. In Special Relativity different observers may disagree about the temporal order of events, but this does not lead to a paradox of effects preceding causes. One event can cause another only if they are close enough in space or distant enough in time so that some signal, traveling at most at the speed of light, can travel from the first to the second. In this case, according to the rules of Special Relativity, all observers will agree that the first event precedes the second. But in quantum mechanics we cannot speak of a signal as having both a definite velocity and a definite starting or finishing point. Hence there is some chance that one event may cause another even if they are so close in time or distant in space that observers may disagree about which event comes first!

This paradox is avoided in modern relativistic quantum mechanics in a subtle way. If one observer sees an electron emitted in the decay of a radioactive nucleus, and then later sees this electron absorbed by a second nucleus, a different observer who sees the electron absorption occur before the electron emission will instead interpret what she sees as the emission of a particle by the second nucleus, followed by its absorption by the first nucleus. Since electric charge is unaffected by motion, both observers would have to agree that the first nucleus loses an electron charge and the second nucleus gains an electron charge; therefore, the observer who sees a particle *emitted* by the second nucleus would report that this particle has an electric charge equal in magnitude but opposite in sign to that of the electron. This particle would not be an electron, but its *antiparticle,* the positron, with the same mass and spin as the electron but a positive electric charge rather than the negative charge of the electron. The paradox of cause following effect can be avoided only if we suppose that for every type of particle there is an antiparticle, with opposite values for electric charge and similar quantities, but in every other way identical to the particle. The mathematical formalism that maintains the relation between particles or antiparticles being emitted and the corresponding antiparticles or particles being absorbed is known as quantum field theory.

Until recently most physicists, including myself, would have said that quantum field theory is the only way of reconciling the peculiar view of time in Special Relativity with the principles of quantum mechanics. Moreover, quantum field theories have been overwhelmingly successful. Everything we have learned about the properties of elementary particles and their weak, strong, and electromagnetic interactions is explained (or we think explainable) in terms of a quantum field theory known as the Standard Model. Only one obstacle remained: there seemed to be no way to incorporate gravitation in any quantum field theory that maintains its consistency at all energies.

This obstacle seems now to have been overcome in relativistic quantum theories of a new sort, known as superstring theories. At the relatively low energies accessible in our laboratories a superstring theory looks like an ordinary quantum field theory, such as the Standard Model of weak, strong, and electromagnetic forces, but at very high energies it is quite different: instead of particles we have strings, tiny one-dimensional vibrating discontinuities in the fabric of spacetime. The different modes of vibration are what we see at low energies as different types of particle.

Once again, it is the problem of time that provides the crucial constraints that make superstring theories what they are. A string as it moves through space sweeps out a two-dimensional surface in spacetime. The vibrations of the string can be described by telling how each point on this surface moves in space and time. There is no problem with the motions in space, but the motions in time, if real, would represent vibrations of negative probability. The only way to avoid this paradox is to impose a symmetry principle, known as conformal symmetry, that allows us to remove these timelike vibrations by a symmetry transformation. The different possible superstring theories simply represent different ways of satisfying the requirements of conformal symmetry.

These ideas have not yet led to a final theory of matter, but they provide what is at present our only real hope of unifying the electroweak, strong, *and* gravitational interactions of the elementary particles. Beyond these challenges, there also remains the old problem of cosmology, the problem of understanding the initial conditions of the universe.

It is here that modern science meets again its old adversary, the chief adversary that Galileo confronted: the preconceptions of philosophy. Many nonscientists (and some scientists) seem to think that there are things that we can know about the nature of time through the exercise of pure reason. Some people think that there must logically have been an origin, a moment before which time did not exist. Biblical fundamentalists may think it was six thousand years ago; the more sophisticated are willing to allow billions of years since the beginning; but they agree on the reality of creation itself. Other people are unable to conceive of an origin to the universe; they take it for granted that every moment must have been preceded by some other moment, so that the universe would have to be infinitely old.

So which argument is right? Neither. The question of the origin of the universe, like the question of the motion of falling bodies, is one that needs to be answered by the methods of science, by theory-aided observation and observation-governed theory, but it is not one that we can settle by pure thought or religious authority.

Through most of this century, the weight of scientific evidence has been in favor of an origin, giving some comfort to those who believe in a supernatural creation. We can trace the present expansion of the matter of the universe backward in time to an epoch when the universe must have been much denser and hotter. The reality of the hot early universe is attested by the discovery of a background of microwave radiation left over from a time, some ten to twenty billion years ago, when the radiation filling the universe was at a temperature of 3,000 K. In the abundance of the various isotopes of the light elements hydrogen, helium, and lithium, we can see further evidence of a time when the temperature of matter and radiation was a billion degrees, a few hundred thousand years earlier than the time when the microwave background was emitted.

Unfortunately, our present theories fail us when we try to apply them at really early times, when the temperature of the universe was 10^{32} (a hundred million million million million million) degrees. At this temperature particles have so much energy that the gravitational forces between them are as strong as any other forces, and matter must be described in the language of superstring theory. No one today knows how to do this.

But we have one piece of evidence that the history of the universe cannot be simply extrapolated back to a moment of infinite temperature a few hundred thousand years before the microwave background was emitted. Looking in different directions in the sky we see remarkable uniformity, especially in the cosmic microwave radiation background. But if the universe had only existed for a few hundred thousand years before the microwave background was emitted, then there would not have been time for any physical influence to go from the source of the radiation in one direction to the source in the opposite direction, and hence no physical mechanism that could have brought these different parts of the universe into equilibrium. This problem has led to the widespread acceptance of the idea of inflation, a long period of expansion before the universe entered into the hot beginning of its present phase. In some modern theories there is no beginning at all, only an endlessly fluctuating universe that here and there occasionally begins a local expansion.

All this is very speculative. We are not even sure that it is appropriate to talk in these terms. In superstring theories it seems that the concept of time itself loses any clear meaning at temperatures of 10^{32} degrees. But we can at least be confident that when these questions are answered (if they ever are answered) they will be answered in the way that we learned from Galileo—through mathematical theories based on observation, and not through the arguments of philosophy or theology.

Galileo was so great a scientist, and his trial and punishment were so dramatic, that it seems to be inevitable for us to try to draw some lesson from his fate. At this meeting I have heard several times that the Church was wrong to condemn Galileo, because religious authority should not intrude on matters that are in the proper sphere of science. Of course I agree with this, but I think that this is only part of the lesson to be learned.

Suppose that Galileo really had been a heretic—that he had held and taught not only views on how the heavens go, but also heretical views on how to go to heaven. Suppose, even worse, that he had been a blasphemer or a public atheist. Under any or all of these circumstances the Church would have been within its rights to denounce his teachings. But it would still have been wrong to threaten him with the instruments of torture or to imprison him. We have today a view of religious and intellectual liberty that condemns the use of force in aid of religious conformity. And this applies whether punishment is imposed by a religious authority like the Inquisition, or a prince like the Grand Duke of Tuscany, or even by a popularly elected government. The overwhelming success of Galileo's scientific achievement makes the injustice of his punishment notorious, but there are millions of others throughout history who have been punished for their views and teachings, and to whom regrets are also due.

I do not say this in any anti-Catholic spirit. In Galileo's time a heretic or an atheist or a blasphemer would not have been safe anywhere in Europe or America, either in Protestant or in Catholic countries, or perhaps anywhere else in the world. And today it is in Islamic countries where religious and intellectual liberty is most in danger, as illustrated by the exercise of criminal jurisdiction by religious courts in Saudi Arabia, and by Iran's despicable sentence of death for blasphemy on Salman Rushdie.

The struggle for religious and intellectual liberty is evidently not over. And in this struggle, the influence of science is surely one of our greatest aids. This is not because of the certainty of scientific knowledge, but precisely because of its uncertainty. Seeing scientists change their minds again and again about matters that are accessible to experiment and mathematical analysis, we are warned against giving the power of criminal jurisdiction to any religious authority that claims to speak with certainty about matters that are beyond human experience.

6

Nature Itself

In 1992 Sir Brian Pippard wrote to ask me to contribute one of the three "Reflections on Twentieth Century Physics" at the close of a new book, *Twentieth Century Physics*, that was being planned by Pippard with Laurie Brown and Abraham Pais. The book was to be a compendium of essays celebrating the successes of physics in the twentieth century. I knew Pippard slightly; in his capacity as holder of the Cavendish Chair at the University of Cambridge (a chair once occupied in succession by Maxwell, Lord Rayleigh, J. J. Thomson, and Ernest Rutherford), he had been my host when I gave the 1975 Scott Lectures at the Cavendish Laboratory. My family and I had had a wonderful visit to Cambridge then, as my lectures had been carefully timed to coincide with the university's annual May Week, a week of parties, dinners, and balls. So I was inclined to accept this invitation for old times' sake. The deciding factor was that one of the other "reflections" would be written by Philip Anderson.

Anderson is a brilliant theorist, one of the leading figures of the past century in the theory of condensed matter physics, the area of physics that studies such properties of matter as superconductivity, semiconductivity, and superfluidity. More to the point, and to my regret, he has also been a leading detractor of the reductionist tendencies of elementary particle physics, as shown for instance in his celebrated essay "More Is Different." And he has been an important opponent of new accelerator construction. I felt that I could not let Anderson have the last word on the past hundred years of physics.

Twentieth Century Physics came out in 1995 in three fat volumes, so heavy that they sat on the desk in my university office for years before I got around to taking them home. When finally

I read Anderson's article I found, along with much good sense, the presentation of an argument against reductionism based on a phenomenon known as "emergence," which Anderson identified as the God Principle of twentieth-century physics. When certain systems reach a sufficient level of complexity, new phenomena "emerge" that do not exist for the elementary constituents of which the system is composed. For instance, superconductivity emerges in solids composed of large numbers of certain atoms, but superconductivity has no meaning for individual atoms. Similarly, life emerges from biochemistry; biochemistry emerges from atomic physics; and atomic physics emerges from the properties of elementary particles as described in the modern Standard Model. The phenomenon of emergence to some extent frees phenomena at large scales from details of what is going on at small scales. For instance, although you have to know a lot about the individual atoms to know whether a given substance is a superconductor, the general features of superconductivity are universal, not dependent on the properties of the atoms composing the superconductor. Anderson is right that further progress in elementary particle physics is unlikely to be of much direct help in understanding emergent phenomena like superconductivity, let alone life.[1]

I don't think that any of this makes a convincing case against the reductionist tendencies of elementary particle physics. After all, emergent phenomena do *emerge*, ultimately from the physics of elementary particles, and if you want to understand why the world is the way it is you have to understand why elementary particles are the way they are. This may not be more important than other types of physics research, but it has a special *kind* of importance, which has led some of us to devote to it our life's work.

The state of science at the end of the twentieth century is very different from its condition at the century's beginning. It is not just

---

1. I say "direct help" because particle physics has provided much indirect help to other sciences through the transfer of technological and intellectual tools, from synchotron radiation to the renormalization group.

that we know more now—we have come in this century to understand the very pattern of scientific knowledge. In 1900 many scientists supposed that physics, chemistry, and biology each operated under its own autonomous laws. The empire of science was believed to consist of many separate commonwealths, at peace with each other, but separately ruled. A few scientists held fast to Newton's dream of a grand synthesis of all the sciences, but without any clear idea of the terms on which this synthesis would be reached. Today we know that chemical phenomena are what they are because of the physical properties of electrons, electromagnetism, and a hundred or so types of atomic nuclei. Biology of course involves historical accidents in a way that physics and chemistry do not, but the mechanism of heredity that drives biological evolution is now understood in molecular terms, and vitalism, the belief in autonomous biological laws, is safely dead. This has truly been the century of the triumph of reductionism.

The same reductionist tendency is visible *within* physics. This is not a matter of how we carry on the practice of physics, but how we view nature itself. There are many fascinating problems that await solution, some, like turbulence, left over from the past, and others recently encountered, like high-temperature superconductivity. These problems have to be addressed in their own terms, not by reduction to elementary particle physics. But when these problems are solved, the solution will take the form of a deduction of the phenomenon from known physical principles, such as the equations of hydrodynamics or of electrodynamics, and when we ask why *these* equations are what they are, we shall trace the answers through many intermediate steps to the same source: the Standard Model of elementary particles. Along with the theory of gravitation and cosmology, the theory of elementary particles thus now constitutes the whole outer frontier of scientific knowledge.

The Standard Model is a quantum field theory. The fundamental ingredients of nature that appear in the underlying equations are fields: the familiar electromagnetic field, and some twenty or so other fields. The so-called elementary particles, like photons and quarks and electrons, are "quanta" of the fields—bundles of the fields' energy and momentum. The properties of these fields and their interactions are largely dictated by principles of symmetry,[2] including Einstein's Special Principle of Relativity, together with a principle of "renormalizability,"[3] which dictates that the fields can interact with each other only in certain specially simple ways. The Standard Model has passed every test that can be imposed with existing experimental facilities.

But the Standard Model is clearly not the end of the story. We do not know why it obeys certain symmetries and not others, or why it contains six types of quarks, and not more or fewer types. Beyond this, appearing in the Standard Model there are about eighteen numerical parameters (like ratios of quark masses) that must be adjusted by hand to make the predictions of the model agree with experiment. Finally, gravitation cannot be brought into the quantum field theoretic framework of the Standard Model, because gravitational interactions do not satisfy the principle of renormalizability that governs the other interactions.

So now we must make the next step toward a truly unified view of nature. Unfortunately, this next step has turned out to be ex-

2. A symmetry principle is a statement that the laws of nature look the same when we change our point of view in some way. The Special Principle of Relativity says that the laws of nature appear the same to observers moving with any constant velocity. There are other spacetime symmetries, which tell us that the laws of nature look the same when we rotate or translate our laboratory, or reset our clocks. The Standard Model is based on a set of symmetry principles, including these symmetries of spacetime, and also other symmetries that require that the laws of nature take the same form when we make certain changes in the ways that the fields of the theory are labeled.

3. The concept of renormalizability arose from the effort in the late 1940s to make sense of the infinite energies and reaction rates that appeared when calculations in quantum electrodynamics were pushed beyond the first approximation. A theory is said to be renormalizable if these infinities can all be canceled by suitable redefinitions (or "renormalizations") of a finite number of parameters of the theory, such as masses and electric charges. Only the simplest theories can be renormalizable. Renormalizability is no longer viewed as a fundamental physical requirement needed to eliminate infinities but, as we shall see, it survives in the Standard Model for other reasons.

traordinarily difficult. We are in a position somewhat like that of Democritus—we can see the outlines of a unifying theory, but it is a theory whose structures become evident only when we examine nature at scales of distance vastly smaller than those accessible in our laboratories. Democritus speculated about atoms, which after over two millennia were discovered to be some ten orders of magnitude smaller (that is, smaller by a factor of one ten-billionth, or 10^{-10}) than Democritus himself. Today we speculate about a theory that would unify all the forces of nature, and we too can see that the structures of this theory must be very much smaller than the scales of distances that we can study experimentally.

We have two clues to the fundamental scale. One is that the strength of the strong nuclear forces (which hold quarks together inside the particles inside the atomic nucleus) decreases very slowly with decreasing distance, while the strengths of the electromagnetic and weak nuclear forces increase even more slowly, all forces becoming equal at a scale of distances that is currently estimated to be about 10^{-16} of the size of an electron.[4] The other clue is that gravitation becomes about as strong as the other forces at distances about 10^{-18} times the size of an electron. These two estimated scales of distance are close enough to encourage us to believe that there really is a final unified theory whose structures become visible at a scale of distances about 10^{-16} to 10^{-18} of the size of an electron. But in a purely numerical sense, we are further away from being able to observe these structures than Democritus was from being able to observe atoms.

So what are we to do? I am convinced that there is no reason for

4. Recall that 10^{-16} is one-tenth multiplied by itself 16 times; that is, it is a decimal point followed by 15 zeros and a one. The electron does not really have a well-defined size; it appears in our theories as a point particle. I am referring here to what is called the classical radius of the electron, that is, the radius that a sphere carrying the charge of the electron would have to have for its electrostatic energy to equal the energy in the electron's mass. This radius can be taken as representative of the characteristic distances encountered in elementary particle physics; for instance, it is not very different from the size of typical atomic nuclei.

us to begin to talk about an end to fundamental physics. Nor do I
think (though I am not sure) that any great change is needed in the
style of physics. There seems to me to be ample hope for progress
in the same reductionist mode that has served us so well in this
century. To be specific, there are two approaches that are not yet
exhausted: a high road, and a low road, with "high" and "low"
here referring to the energy of the processes studied.

One approach, the low road, is to try through the combined
efforts of theorists and experimentalists to complete our understanding of physics at energies accessible in our laboratories. The
higher the energy, the smaller the scale of the structures that can be
studied; with energies no larger than about a trillion electron volts,
we cannot directly study structures smaller than about one ten-
thousandth the size of the electron. As already mentioned, physics
on this scale is known to be well described by the Standard Model.
This theory is based in part on a principle of symmetry that pro-
hibits the appearance in the equations of the theory of any masses
for the known elementary particles. Something beyond the known
particles must break this symmetry. In the original version of the
theory, this something was a field pervading the universe, which
breaks the symmetry in much the same way that the gravitational
field of the earth breaks the symmetry between up and down. The
quanta of this field would show up in our experiments as a particle, known as the Higgs particle. There are other alternatives, but
all theories agree that something beyond the known particles of
the Standard Model must show up when we study collisions at energies high enough to create masses of the order of a trillion electron volts, a thousand times the mass of the proton.

It is vitally important to learn the details of the symmetry-breaking mechanism, because it is this mechanism that sets the scale of
the masses of the quarks and the electron and other known elementary particles; the only mass appearing in the equations of the
Standard Model in its simplest version is the Higgs particle mass.
From units of mass we can infer units of length, so this is also what
sets the scale of lengths characteristic of the elementary particles.

Thus we must understand this symmetry-breaking mechanism if we are to solve the "hierarchy problem," the problem of making sense of the enormous disparity between the lengths encountered in our laboratory, quantities like the size of the electron, and the scale of distances where all the forces become unified.

It was this question that we hoped to settle at the Superconducting Super Collider. Now that the Congress of the United States has decided to cancel the Super Collider, our hopes are riding on plans to build a similar accelerator, the Large Hadron Collider, in Europe. In the meanwhile, we have a good chance of getting advance information about where the Higgs particle or whatever takes its place might be found. The point is that quantum fluctuations involving the continuous creation and annihilation of this particle (or particles) have a slight effect on quantities that can be measured with present facilities, quantities like the mass of the W particle. We can expect the mass of the W particle to become much more accurately known through experiments at CERN, while experiments at Fermilab near Chicago are now in the process of providing an accurate value for the mass of the "top" quark, which also affects the mass of the W particle. When this is done, we will be able to estimate the mass of the Higgs particle (if there is just one), or rule out the possibility of a single Higgs particle at any mass. But we will still need the Large Hadron Collider to pin down the details of the symmetry-breaking mechanism that gives mass to the elementary particles.

There are many other things that might turn up in our experiments while we are waiting for the next generation of large accelerators. These include an electric dipole moment (like the magnetization of a compass needle, but producing electric rather than magnetic fields) of the neutron or electron, an effect that is expected in some theories of symmetry breaking. We also might discover any of the new particles predicted by various theories (supersymmetry, technicolor) that have been suggested in attempts to solve the hierarchy problem. There are also tiny effects like proton decay or neutrino masses that might be directly produced by the physical processes that occur at the very short distance scales of the final unification. These exotic effects would be manifested as additions to the Standard Model that violate the principle of renormalizability mentioned earlier. Though not forbidden, these effects would be proportional to powers of the very small distances from which they arise, and hence they would be very small, which is presumably why they have not yet been detected. But they may yet be. There will soon be a new underground facility in Japan known as Kamiokande, where proton decay might be discovered, and there are indirect hints from studies of neutrinos from the sun that neutrinos actually do have small masses. Any or all of these things might be discovered in future experiments. But we have been saying this for almost twenty years, since the completion of the Standard Model, and so far none of them has turned up.

Then there is the high road. A fair fraction of particle theorists are now trying to jump over all intermediate steps and proceed directly to the formulation of a final unified theory, without waiting for new data. For a while in the 1970s, after the success of the Standard Model, it was thought that this final theory might take the form of a quantum field theory, like the Standard Model, but simpler and more unified. This hope has been largely abandoned. For one thing, we now understand that *any* physically satisfactory theory will look like a quantum field theory at sufficiently low energy, so the great success of quantum field theory in the Standard Model tells us little about the deeper theory from which it derives. The Standard Model is seen as an "effective" quantum field theory—the low energy limit of a quite different fundamental theory. Also, the continued failure to bring gravitation into the ambit of quantum field theory suggests that the fundamental theory is quite different.

The best hope for success along this line is in some sort of string theory. Strings are hypothetical one-dimensional elementary entities, either closed like rubber bands or open like pieces of ordinary string, which like a violin string can vibrate at a great many different frequencies. They are about 10^{-18} times the size of an electron, so when viewed in our laboratories they seem like point particles of various different types, the type depending on the mode in which the string happens to be vibrating. One of the exciting things about string theories is that in some versions they predict a menu of particle types that is impressively similar to what we actually observe in nature. Further, one of the particle types that is predicted by all versions of string theory is the graviton, the quantum of the gravitation field. Thus string theories not only unite gravitation with the rest of elementary particle physics, they explain why gravitation must exist. Finally, the old problem of infinities that stood in the way of a quantum field theory of gravitation is avoided in string theories. The infinities found in our field-theoretic calculations arose from processes in which particles occupy the same point in space. But strings have a finite extension, and therefore a given point on one string will always be at a non-zero distance from almost all points on any other string. String theory thus provides our first real candidate for a final unifying theory.

String theory unfortunately has not yet lived up to the great expectations that attended it in the 1980s. There seem to be a large variety of different string theories, and although it is widely suspected that these are really just different versions of the same universal theory, no one knows what that theory might be. There are formidable mathematical obstacles that stand in the way of solving any of the string theories to find just what it predicts for measurable physical quantities like quark masses, and we cannot tell from first principles which version of string theory is correct. Even if all these obstacles are overcome, we will still be left with the question why the real world *should* be described by anything like string theory.

One possible answer to this question is suggested by reference to the historical origins of string theory. In the 1960s, before the advent of the now standard quantum field theory of the strong nuclear forces, many theorists had given up on the idea of describing these forces in terms of any quantum field theory. Instead they sought to calculate the properties of nuclear particles and mesons through a positivistic program, known as "S-matrix theory," that avoids referring to unobservable quantities like the field of the electron. In this program one proceeds by imposing physically reasonable conditions on observable quantities, specifically on the probabilities of all possible reactions among any numbers of particles. (Among these conditions was the condition that the probabilities of all reactions would always add up to 100 percent, another requirement that these probabilities depend in a smooth way on the energies and directions of motion of the particles participating in any reaction, yet another requirement regarding the behavior of these probabilities at very high energy, and finally various symmetry conditions, including the symmetries of spacetime embodied in the Special Principle of Relativity.) It turned out to be extraordinarily difficult to find *any* set of probabilities that satisfied all of these conditions. Finally, by inspired guesswork, a formula for reaction probabilities was found in 1968–69 that seemed to satisfy all these conditions. Shortly after, it was realized that the theory that had been discovered was in fact a theory of strings. It is possible that this history reflects the logical basis of string theory. That is, string theory may ultimately be understood as the only way to satisfy all of the physically reasonable conditions on reaction probabilities, or at least the only way to satisfy these conditions in any theory that includes gravitation.

This point of view has its paradoxical elements. When in S-matrix theory we talk of the probabilities of various "reactions," we have in mind processes in which two or more particles come together after traveling freely over great distances, and then interact, producing new particles that finally travel apart until they are again so far apart that they can no longer interact. This is the paradigmatic experiment of modern elementary particle physics. But this sort of reaction can occur only in a universe that is more or less empty and "flat"—that is, not filled with a high density of matter, and not pervaded by strong gravitational fields that warp the structure of spacetime. This is indeed the present state of our universe, but things were not like that in the early universe, and even today there are objects like black holes where spacetime is grossly curved. It seems odd to take a set of "reasonable conditions" on reaction probabilities as the fundamental principles of physics, when the reactions to which these conditions refer were not always possible, and are not everywhere possible even today.

Indeed, the fact that our present universe is more or less empty and flat is itself rather paradoxical. In most theories, the quantum fluctuations of various fields would give "empty" space an energy density so great that the gravitational field it produces would make spacetime grossly curved—so much so, that nothing like an ordinary elementary particle reaction could occur, and no scientists could live to observe it.[5] This problem is not solved in string theory; most of the large number of proposed string theories predict an enormous vacuum energy density.

To solve the problem of the vacuum energy density, we may need to invoke physical principles that are not only new, but of a new type, different from those that have so far seemed legitimate. This would not be the first time that there have been changes in our ideas of what are permissible fundamental principles. In his 1909 book *The Theory of the Electron*, Hendrik Lorentz took the opportunity to comment on the difference between Einstein's Special Theory of Relativity, proposed four years earlier, and his own work. Lorentz had tried to use an electromagnetic theory of electron structure to show that matter composed of electrons would when in motion behave in such a way as to make it impossible to detect the effects of its motion on the speed of light, thus explaining the persistent failure to detect any difference in the speed of light along or at right angles to the earth's motion around the sun. Einstein in contrast had taken as a fundamental axiom that the speed of light is the same to all observers. Lorentz grumbled that "Einstein simply postulates what we have deduced, with some dif-

5. For historical reasons, this is known as the problem of the cosmological constant.

ficulty and not always satisfactorily, from the fundamental equations of the electromagnetic field." But history was on Einstein's side. From a modern point of view, what Einstein had done was to introduce a principle of symmetry—the invariance of the laws of nature under changes in the velocity of the observer—as one of the fundamental laws of nature. Since Einstein's time, we have become more and more familiar with the idea that symmetry principles of various sorts are legitimate fundamental hypotheses. The Standard Model is largely based on a set of assumed symmetry principles, and string theory can also be viewed in this way. But in the days of Einstein and Lorentz, symmetries were generally regarded as mathematical curiosities, of great value to crystallographers, but hardly worthy to be included among the fundamental laws of nature. It is not surprising that Lorentz was uncomfortable with Einstein's hypothesis of the Principle of Relativity. We too may have to discover new sorts of hypotheses, which may at first seem to us as uncongenial as Einstein's symmetry principle seemed to Lorentz.

We have already encountered such a hypothesis. The so-called anthropic principle states that the laws of nature must allow for the appearance of living beings capable of studying the laws of nature. This principle is certainly not widely accepted today, though it provides what is so far the only way we have of solving the problem of a large vacuum energy density. (Too large a vacuum energy density would, depending on its sign, either prevent galaxies from forming, or end the Big Bang too early for life to evolve.) In some cosmological theories, a weak version of the anthropic principle would be no more than common sense. If what we now call the laws of nature, including the values of physical constants, vary from place to place in the universe, or from one epoch to another, or from one term in the quantum mechanical wave function of the universe to another, then naturally we could only find ourselves in a place or an epoch or a term in the wave function hospitable to intelligent life.

In the long run we may have to invoke a stronger form of the anthropic principle. When at last we learn the final laws of physics, we will then confront the question of why nature is described by these laws, and none other. There are plenty of imaginable theories that are logically perfectly consistent but nonetheless wrong. For instance, there is nothing *logically* wrong with Newtonian mechanics. Conceivably, the correct final theory is the only logically consistent set of principles consistent with the appearance of intelligent life.

The anthropic principle is just one example of an unconventional hypothesis, but one that has already received serious attention. In our search for the final laws of nature, we may in the twenty-first century have to accept the legitimacy of new sorts of fundamental physical principles stranger than any that have so far been imagined.

7

The Boundaries
of Scientific Knowledge

———

In 1994 I was invited by *Scientific American* to contribute the "keynote" article to an issue celebrating the one hundred and fiftieth anniversary of the magazine. I had written two earlier articles for *Scientific American,* on the unification of weak and electromagnetic forces and on the possible decay of the proton, and was not keen to write another just then, but sesquicentennials don't come along too often, so I accepted. I was led to think that the issue would assess the present state of scientific knowledge, and I dutifully wrote an essay about the limits of this knowledge. As it happened, while I was working on the article the title of the issue was shifted to "Life in the Universe." The reader will see in this article some weak last-minute attempts to tie in to that topic, but unfortunately my article remained only remotely relevant to life in the universe. But it was still the keynote article, so the editors gave it the same title as the issue. They also dropped the connection with their one hundred and fiftieth anniversary. It was not an entirely happy experience, but at least now I have the pleasure of seeing my article appearing here under its original title.

———

In Walt Whitman's often quoted poem "When I Heard the Learn'd Astronomer," the poet tells how, being shown the astronomer's charts and diagrams, he became tired and sick and wandered off by himself to look up "in perfect silence at the stars." Generations of scientists have been annoyed by these lines. The sense of beauty and wonder has not atrophied through the work of science, as

Whitman implies. The night sky is as beautiful as ever, to astronomers as well as to poets. And as we understand more and more about nature, the scientist's sense of wonder has not diminished but has rather become sharper, more narrowly focused on the mysteries that still remain.

The nearby stars that Whitman could see without a telescope are now not so mysterious. Massive computer codes simulate the nuclear reactions at the stars' cores and follow the flow of energy by convection and radiation to their visible surfaces, explaining both their present appearance and how they have evolved. The observation in 1987 of gamma rays and neutrinos from a supernova in the Large Magellanic Cloud provided dramatic confirmation of the theory of stellar structure and evolution. These theories are themselves beautiful to us, and knowing why Betelgeuse is red may even add to the pleasure of looking at the winter sky.

But there are plenty of mysteries left, many of them discussed by other authors in this issue. Of what kind of matter are galaxies and galactic clusters made? How did the stars, planets, and galaxies form? How widespread in the universe are habitats suitable for life? How did the earth's oceans and atmosphere form? How did life start? What are the relations of cause and effect between the evolution of life and the terrestrial environment in which it has occurred? How large is the role of chance in the origin of the human species? How does the brain think? How do human institutions respond to environmental and technological change?

We may be very far from the solution of some of these problems. Still, we can guess what kinds of solutions they will have, in a way that was not possible when *Scientific American* was founded a hundred and fifty years ago. New ideas and insights will be needed, which we can expect to find within the boundaries of science as we know it.

Then there are mysteries at the outer boundaries of our science, matters that we cannot hope to explain in terms of what we already know. When we explain any observation, it is in terms of scientific principles that are themselves explained in terms of deeper principles. Following this chain of explanations, we are led at last to laws of nature that cannot be explained within the boundaries of contemporary science.

Also, in dealing with life and many other aspects of nature, our explanations have a historical component. Some historical facts are accidents that can never be explained, except perhaps statistically: we can never explain precisely why life on the earth takes the form it does, although we can hope to show that some forms are more likely than others.

We can explain a great deal, even where history plays a role, in terms of the conditions with which the universe began, as well as the laws of nature. But how do we explain the initial conditions? A further complex of puzzles overhangs the laws of nature and the initial conditions. It concerns the dual role of intelligent life—as part of the universe we seek to explain, and as the explainer.

The laws of nature as we currently understand them allow us to trace the observed expansion of the universe almost all the way back to what would be a true beginning, a moment when the universe was infinitely hot and dense, some ten to twenty billion years ago. We do not have enough confidence in the applicability of these laws at extreme temperatures and densities to be sure that there really was such a moment, much less to work out all the initial conditions, if there were any. For the present, we cannot do better than to describe the initial conditions of the universe at a time about 10^{-12} second after the nominal moment of infinite temperature.[1]

The temperature of the universe had dropped by then to about 10^{15} degrees, cool enough for us to apply our present physical theories. At these temperatures the universe would have been filled with a gas consisting of all the types of particles known to high energy nuclear physics, together with their antiparticles, continually being annihilated and created in their collisions. As the universe

1. Here again I use a notation for very large or small numbers in terms of powers of 10 or 1/10. For instance, 10^{-12} is 1/10 multiplied by itself 12 times, or 0.000000000001, while 10^{16} is 10 multiplied by itself 16 times. [Added note.]

continued to expand and cool, creation became slower than anni-hilation, and almost all the particles and antiparticles disappeared. If there had not been a small excess of electrons over antielectrons, and quarks over antiquarks, then ordinary particles like electrons and quarks would be virtually absent in the universe today. It is this early excess of matter over antimatter, estimated as one part in about 10^{10}, that survived to form light atomic nuclei three minutes later, then after a million years to form atoms and later to be cooked to heavier elements in stars, ultimately to provide the ma-terial out of which life would arise. The one part in 10^{10} excess of matter over antimatter is one of the key initial conditions that de-termined the future development of the universe.

In addition, there may exist other types of particles, not yet ob-served in our laboratories, that interact more weakly with one an-other than do quarks and electrons and that therefore would have annihilated relatively slowly. Large numbers of these exotic parti-cles would have been left over from the early universe, forming the "dark matter" that now apparently makes up much of the mass of the universe.

Finally, although it is generally assumed that when the universe was 10^{-12} second old its contents were pretty nearly the same ev-erywhere, small inhomogeneities must have existed that triggered the formation, millions of years later, of the first galaxies and stars. We cannot directly observe any inhomogeneities at times earlier than about a million years after the beginning, when the universe first became transparent. Astronomers are currently engaged in mapping minute variations in the intensity of the cosmic micro-wave radiation background that was emitted at that time, using them to infer the primordial distribution of matter. This informa-tion can in turn be used to deduce the initial inhomogeneities at 10^{-12} second after the beginning.

From the austere viewpoint of fundamental physics, the history of the universe is just an illustrative example of the laws of nature. At the deepest level to which we have been able to trace our expla-nations, those laws take the form of quantum field theories. When quantum mechanics is applied to a field such as the electromagnetic field, it is found that the energy and momentum of the field come in bundles, or quanta, that are observed in the laboratory as particles. The modern Standard Model posits an electromagnetic field, whose quanta are photons; an electron field, whose quanta are electrons and antielectrons; and a number of other fields whose quanta are particles called leptons and antileptons. There are various quark fields, whose quanta are quarks and antiquarks, and there are eleven other fields whose quanta are the particles that transmit the weak and strong forces that act on the elementary particles.

The Standard Model is certainly not the final law of nature. Even in its simplest form it contains a number of arbitrary features. Some eighteen numerical parameters appear in the theory, with values that have to be taken from experiments, with little understanding why these values are what they are. The multiplicity of types of quarks and leptons is also unexplained. Further, one aspect of the model is still uncertain: we are not sure of the details of the mechanism that gives masses to the quarks, electrons, and other particles. This is the puzzle that was to have been solved by the now canceled Superconducting Super Collider. We hope it will be unraveled by the Large Hadron Collider being planned at CERN, near Geneva. Finally, the model is incomplete; it does not include gravitation. We have a good field theory of gravitation, the General Theory of Relativity, but the quantum version of this theory breaks down at very high energies.

It is possible that all these problems will find their solution in a new kind of theory known as string theory. The point particles of quantum field theory are reinterpreted in string theory as tiny, extended one-dimensional objects called strings. These strings can exist in various modes of vibration, each mode appearing in the laboratory as a different type of particle. String theory does not merely provide a quantum description of gravitation that makes sense at all energies: one of the modes of vibration of a string would appear as a particle with the properties of the graviton, the quantum of the gravitational field, so string theory even offers an explanation of why gravitation exists. Further, there are versions of string theory that predict something like the menu of fields incorporated in the Standard Model.

But string theory has had no successes yet in explaining or predicting any of the numerical parameters of the Standard Model. Moreover, strings are much too small for us to detect directly the stringy nature of elementary particles; a string is smaller relative to an atomic nucleus than is a nucleus relative to a mountain. The intellectual investment now being made in string theory without the slightest encouragement from experimentation is unprecedented in the history of science. Yet for now, it offers our best hope for a deeper understanding of the laws of nature.

The present gaps in our knowledge of the laws of nature stand in the way of explaining the initial conditions of the universe, at 10^{-12} second after the nominal beginning, in terms of the history of the universe at earlier times. Calculations in the past few years have made it seem likely that the tiny excess of quarks and electrons over antiquarks and antielectrons at this time was produced a little earlier, at a temperature of about 10^{16} degrees. At that moment the universe went through a phase transition, something like the freezing of water, in which the known elementary particles for the first time acquired mass. But we cannot explain why the excess produced in this way should be about one part in 10^{10}, much less calculate its precise value, until we understand the details of the mass-producing mechanism.

The other initial condition, the degree of inhomogeneity in the early universe, may trace back to even earlier times. In our quantum field theories of elementary particles, including the simplest version of the Standard Model, several fields pervade the universe, taking non-zero values even in supposedly empty space. In the present state of the universe, these fields have reached equilibrium values, which minimize the energy density of the vacuum. This vacuum energy density, also known as the cosmological constant, can be measured through the gravitational field that it produces. It is apparently very small.

In some modern theories of the early universe, however, there was a very early time when these fields had not yet reached their equilibrium values, so that the vacuum would have had an enormous energy density. This energy would have produced a rapid expansion of the universe, known as inflation. Tiny inhomogeneities that would have been produced by quantum fluctuations during this inflation would have been magnified in the expansion and could have produced the much larger inhomogeneities that millions of years later triggered the formation of galaxies.

It has even been conjectured that the inflation that began the expansion of the visible universe did not occur throughout the cosmos. It may instead have been just one local episode in an eternal succession of local inflations that occur at random throughout an infinite universe. If this is true, then the problem of initial conditions disappears; there was no initial moment.

In this picture, our local expansion may have begun with some special ingredients or inhomogeneities, but like the forms of life on the earth, these could be understood only in a statistical sense. Unfortunately, at the time of inflation gravitation was so strong that quantum gravitational effects were important. So these ideas will remain speculative until we understand the quantum theory of gravitation—perhaps in terms of something like a string theory.

The experience of the past hundred and fifty years has shown that life is subject to the same laws of nature as is inanimate matter. Nor is there any evidence of a grand design in the origin or evolution of life. There are well-known problems in the description of consciousness in terms of the working of the brain. They arise because we each have special knowledge of our own consciousness that does not come to us from the senses. But I don't think that this means that consciousness will never be explained. The fundamental difficulties in understanding consciousness do not stand in the way of explaining the *behavior* of other people in terms of neurology and physiology and, ultimately, in terms of physics and history. When we have succeeded in this task, we will doubtless find that part of the explanation of behavior is a program of neural activity, that we will recognize as corresponding to our own consciousness.

Much as we would like to take a unified view of nature, we keep encountering a stubborn duality in the role of intelligent life in the universe, as both the observer of nature and part of what is observed. We see this even at the deepest level of modern physics. In quantum mechanics the state of any system is described by a mathematical object known as the wave function. According to the interpretation of quantum mechanics worked out in Copenhagen in the early 1930s, the rules for calculating the wave function are of a very different character from the principles used to interpret it. On the one hand, there is the Schrödinger equation, which describes in a perfectly deterministic way how the wave function of any system changes with time. Then, quite separate, there is a set of principles that tells how to use the wave function to calculate the probabilities of various possible outcomes when someone makes a measurement.

The Copenhagen interpretation holds that when we measure any quantity, such as position or momentum, we are intervening in a way that causes an unpredictable change in the wave function, resulting in a new wave function for which the measured quantity has some definite value, in a manner that cannot be described by the deterministic Schrödinger equation. For instance, before a measurement the wave function of a spinning electron is generally a sum of terms corresponding to different directions of the electron's spin; with such a wave function the electron cannot be said to be spinning in any particular direction. If we measure whether the electron is spinning clockwise or counterclockwise around some axis, however, we somehow change the electron's wave function so that the electron is definitely spinning one way or the other. Measurement is thus regarded as something intrinsically different from anything else in nature. And although opinions differ, it is hard to identify anything special that qualifies some process to be called a measurement, except its effect on a conscious mind.

Among physicists and philosophers one finds at least four different reactions to the Copenhagen interpretation. The first is simply to accept it as it stands. This attitude is mostly limited to those who are attracted to the old, dualistic world view that puts life and consciousness on a different footing from the rest of nature. The second attitude is to accept the rules of the Copenhagen interpretation for practical purposes, without worrying about their ultimate interpretation. This attitude is by far the most common among working physicists. The third approach is to try to avoid these problems by changing quantum mechanics in some way. So far no such attempt has found much acceptance among physicists.

The final approach is to take the Schrödinger equation seriously, to give up the dualism of the Copenhagen interpretation, and to try to explain its successful rules through a description of measurers and their apparatus in terms of the same deterministic evolution of the wave function that governs everything else. When we measure some quantity (like the direction of an electron's spin), we put the system in an environment (for instance, a magnetic field) where its energy (or its trajectory) has a strong dependence on the value of the measured quantity. According to the Schrödinger equation, the different terms in the wave function that correspond to different energies will oscillate rapidly at rates proportional to these energies.

A measurement thus makes the terms of the wave function that correspond to different values of a measured quantity, such as an electron's spin, oscillate rapidly at different rates, so that they cannot interfere with one another in any future measurement, just as the signals from radio stations broadcasting at widely spaced frequencies do not interfere. In this way, a measurement causes the history of the universe for practical purposes to diverge into different, noninterfering tracks, one for each possible value of the measured quantity.

Yet how do we explain the Copenhagen rules for calculating the probabilities for these different "worldtracks" in a world governed by the completely deterministic Schrödinger equation? Progress has recently been made on this problem, but it is not yet definitely solved. (For what it is worth, I prefer this last approach, although the second has much to recommend it.)

It is also difficult to avoid talking about living observers when we ask why our physical principles are what they are. Modern quantum field theory and string theory can be understood as answers to the problem of reconciling quantum mechanics and Special Relativity in such a way that experiments are guaranteed to give sensible results. We require that the results of our dynamical calculations must satisfy conditions that are known to field theorists as unitarity, positivity, and cluster decomposition. Roughly speaking, these conditions require that probabilities always add up to 100 percent, that they are always positive, and that probabilities measured in distant experiments are not related.

This is not so easy. If we try to write down some dynamical equations that will automatically give results consistent with some of these conditions, we usually find that the results violate the other conditions. It seems that any relativistic quantum theory that satisfies all these conditions must appear at sufficiently low energy like a quantum field theory. That is presumably why nature at accessible energies is so well described by the quantum field theory known as the Standard Model.

Also, so far as we can tell, the only mathematically consistent relativistic quantum theories that satisfy these conditions at all energies and that involve gravitation are string theories. Further, the student of string theory who asks why one makes this or that mathematical assumption is told that otherwise one would violate physical principles like unitarity and positivity. But why are these the correct conditions to impose on the results of all imaginable experiments if the laws of nature allow the possibility of a universe that contains no living beings to carry out experiments?

This question does not intrude on much of the actual work of theoretical physics, but it becomes urgent when we seek to apply quantum mechanics to the whole universe. At present, we do not understand even in principle how to calculate or interpret the wave function of the universe, and we cannot resolve these problems by requiring that all experiments give sensible results, because by definition there is no observer outside the universe who can experiment on it.

These mysteries are heightened when we reflect how surprising it is that the laws of nature and the initial conditions of the universe should allow for the existence of beings who could observe it. Some have argued that life as we know it would be impossible if any one of several physical quantities had slightly different values. The best known of these quantities is the energy of one of the excited states of the carbon 12 nucleus. There is an essential step in the chain of nuclear reactions that build up heavy elements in stars. In this step, two helium nuclei join together to form the unstable nucleus of beryllium 8, which sometimes before fissioning absorbs another helium nucleus, forming carbon 12 in this excited state. The carbon 12 nucleus then emits a photon and decays into the stable state of lowest energy. In subsequent nuclear reactions carbon is built up into oxygen and nitrogen and the other heavy elements necessary for life. But the capture of helium by beryllium 8 is a resonant process, whose reaction rate is a sharply peaked function of the energies of the nuclei involved. If the energy of the excited state of carbon 12 were just a little higher, the rate of its formation would be much less, so that almost all the beryllium 8 nuclei would fission into helium nuclei before carbon could be formed. The universe would then consist almost entirely of hydrogen and helium, without the ingredients for life.

Opinions differ as to the degree to which the constants of nature must be fine-tuned to make life necessary. There are independent reasons to expect an excited state of carbon 12 near the resonant energy. But one constant does seem to require an incredible fine-tuning: it is the vacuum energy, or cosmological constant, that I mentioned in connection with inflationary cosmologies.

Although we cannot calculate this quantity, we can calculate some contributions to it (such as the energy of quantum fluctuations in the gravitational field that have wavelengths no shorter than about 10^{-33} centimeter, where our present theory of gravitation becomes inapplicable). These contributions come out about 10^{120} times larger than the maximum value allowed by our observations of the present rate of cosmic expansion. If the various contributions to the vacuum energy did not nearly cancel, then, depending on whether the total vacuum energy is negative or positive, the universe either would go through a complete cycle of expansion and contraction before life could arise or would expand so rapidly that no galaxies or stars could form.

Thus the existence of life of any kind seems to require a cancellation between different contributions to the vacuum energy, accurate to about 120 decimal places. It is possible that this cancellation will be explained in terms of some future theory. So far, in string theory as well as in quantum field theory, the vacuum energy involves arbitrary constants, which must be carefully adjusted to make the total vacuum energy small enough for life to be possible.

All these problems may eventually be solved without supposing that life or consciousness plays any special role in the fundamental laws of nature or initial conditions. It may be that what we now call the constants of nature actually vary from one part of the universe to another. (Here "different parts of the universe" could be understood in various senses. The phrase could, for example, refer to different local expansions arising from episodes of inflation in which the fields pervading the universe took different values, or else to the different quantum mechanical worldtracks that arise in some versions of quantum cosmology.) If this is the case, then it would not be surprising to find that life is possible in some parts of the universe, though probably not in most. Naturally, any living beings who evolve to the point where they can measure the constants of nature will always find that these constants have values that allow life to exist. The constants have other values in other parts of the universe, but there is no one there to measure them. (This is one version of what is sometimes called the anthropic principle.) Still, this presumption would not indicate any special role for life in the fundamental laws, any more than the fact that the sun has a planet on which life is possible indicates that life played a role in the origin of the solar system. The fundamental laws would be those that describe the *distribution* of values of the constants of nature between different parts of the universe, and in these laws life would play no special role.

If the content of science is ultimately impersonal, its conduct is part of human culture, and not the least interesting part. Some philosophers and sociologists have gone so far as to claim that scientific principles are, in whole or in part, social constructions, like the rules of contract law or contract bridge. Most working scientists find this "social constructivist" point of view inconsistent with their own experience. Still, there is no doubt that the social context of science has become increasingly important to scientists, as we need to ask society to provide us with more and more expensive tools: accelerators, space vehicles, neutron sources, genome projects, and so on.

It does not help that some politicians and journalists assume that the public is interested only in those aspects of science that promise immediate practical benefits to technology or medicine. Some work on the most interesting problems of biological or physical science does have obvious practical value, but some does not, especially research that addresses problems lying at the boundaries of scientific knowledge. To earn society's support, we have to make true what we often claim: that today's basic scientific research is part of the culture of our times.

Whatever barriers now exist to communication between scientists and the public, they are not impermeable. Isaac Newton's *Principia* could at first be understood only by a handful of Europeans. Then the news that we and our universe are governed by precise, knowable laws did eventually diffuse throughout the civilized world. The theory of evolution was strenuously opposed at first; now creationists are an increasingly isolated minority. Today's research at the boundaries of science explores environments of energy and time and distance far removed from those of everyday life and often can be described only in esoteric mathematical language. But in the long run, what we learn about why the world is the way it is will become part of everyone's intellectual heritage.

8

The Methods of Science . . .
and Those by Which We Live

The National Association of Scholars is an organization of academics, united chiefly by a shared allegiance to a traditional view of scholarship as a search for objective truth. The fifth national conference of this association, titled "Objectivity and Truth in the Natural Sciences, the Social Sciences, and the Humanities," met in Cambridge, Massachusetts, in November 1994. I gave the opening talk, which is reproduced below. I thought it would be preaching to the choir, but I encountered some disagreement after my talk, which is described in the next essay in this collection.

I also received an interesting comment on the published version of the talk. I had mentioned that my cat could not recognize a photo of a fish, and someone wrote me to say that if my cat was that unintelligent then he must be a Siamese. I would have thought it was a silly remark, except that in fact my cat *is* a Siamese.

Standing in a bookshop in Harvard Square a decade ago, I noticed a book on the philosophy of science by a friend of mine. I opened it and found it interesting—my friend was discussing questions about scientific knowledge that I had not thought of asking. Yet, although I bought the book, I realized that it would probably be a long time before I sat down to read it, because I was busy with my research and I knew deep in my heart that this book was not going to help me in my work as a scientist.

This was not a great discovery. I think few philosophers of science take it as part of their job description to help scientists in their research. Ludwig Wittgenstein and others have explicitly disclaimed any such aim. But then, standing in the bookshop, it occurred to me to ask why this should be? Why should the philosophy of science not be of more help to scientists? I raise this question here not in order to attack the philosophy of science, but because I think that it is an interesting question—perhaps even philosophically interesting.

In thinking this over, I realized that a good part of the explanation must be that science is a moving target, that the standards for successful scientific theories shift with time. It is not just our view of the universe that shifts, but our view of what kinds of views we should have or can have. How can we expect a philosopher (or anyone else) to know enough about the universe and the human mind to anticipate these shifts in advance? How could anyone expect Aristotle in his writings on projectiles to imagine the quantitative study of the motion of projectiles that began just a generation later in Alexandria? To Aristotle, although he knew about quantitative sciences like arithmetic and astronomy, the sciences appropriate for most of nature were little more than taxonomy. We learn about the philosophy of science by doing science, not the other way around.

A favorite example of mine, one much closer to home, is presented by Albert Einstein's development of the Special Theory of Relativity in 1905. For some years before 1905 a number of physicists had been worrying about why it seemed to be impossible to detect any effect on the speed of light of the earth's motion through the ether. The electron, just discovered in 1897, was then the only known elementary particle, and it was widely supposed that all matter was composed of electrons. So most physicists who worried about the ether, such as Abraham, Lorentz, and Poincaré, tried to develop a theory of the electron's structure that would describe how the lengths of measuring rods and the rates of clocks made of electrons change as they move through the ether in just such a way as to make it seem that the speed of light does not depend on the speed of the observer.

Einstein did not proceed along that line. Instead, he took as a fundamental hypothesis the principle of relativity, that it is not possible to detect the effects of uniform motion on the speed of light or anything else. On this foundation, he built a whole new theory of mechanics. In this work Einstein set the tone of the twentieth century by taking a principle of symmetry, or invariance—a principle that says that some changes in point of view cannot be detected—as a fundamental part of scientific knowledge, a hypothesis at the very roots of science, rather than something that is unsatisfactory until it can be deduced from a specific dynamical theory. In other words, Einstein had changed the way that we score our theories.

Not only does the fact that the standards of scientific success shift with time make the philosophy of science difficult; it also raises problems for the public understanding of science. We do not have a fixed scientific method to rally round and defend. I remember a conversation I had years ago with a high school teacher, who explained proudly that in her school teachers were trying to get away from teaching just scientific facts, and wanted instead to give their students an idea of what the scientific method was. I replied that I had no idea what the scientific method was, and I thought she ought to teach her students scientific facts. She thought I was just being surly. But it's true; most scientists have very little idea of what the scientific method is, just as most bicyclists have very little idea of how bicycles stay erect. In both cases, if they think about it too much, they're likely to fall off.

The changes in the way we judge our theories have bothered philosophers and historians of science. Thomas Kuhn's early book, *The Structure of Scientific Revolutions*, emphasized this process of change in our scientific standards. I think Kuhn went overboard in concluding that there was a complete incommensurability between present and past standards, but it is correct that there is a qualitative change in the kind of scientific theory we want to develop that has taken place at various times in the history of science. But Kuhn then proceeded to the fallacy—much clearer in what he has written recently—that in science we are not in fact moving toward objective truth. I call this a fallacy because it seems to me a simple non sequitur. I do not see why the fact that we are discovering not only the laws of nature in detail, but what kinds of laws are worth discovering, should mean that we are not making objective progress.

Of course, it's hard to prove that we are making objective progress. David Hume showed early on the impossibility of using rational argument to justify the scientific method, since rational argument, that is, appeal to experience, is part of the scientific method. But this kind of skepticism gets one nowhere. One can be equally skeptical about our knowledge of ordinary objects, because the methods of science are not that different, except in degree, from the methods by which we live our lives. But in fact we don't worry very much about whether our knowledge of common objects like chairs is objective or socially constructed.

The physicist who lives with principles of symmetry, such as the principles of relativity, or with more esoteric constructs like quarks or quantum fields or superstrings, gets nearly as familiar with them as with the chairs on which he or she sits. The physicist finds that, just as with chairs, these constructs can't be made up as one goes along, that they seem to have an existence of their own. If you say the wrong thing about them, you'll find out about it pretty soon, when experiment or mathematical demonstration proves that you've been wrong.

My experience with the principles of physics is not that different from my experience of chairs. Even with chairs one can raise the question of different levels of knowledge. We mostly know about chairs by sitting in them or by bumping into them, but there are other ways of knowing about chairs. For instance, you can look at photographs of chairs. That's a mode of knowing about chairs that's actually fairly refined. I believe there are primitive peoples—though I'm not sure about this—who do not recognize photographs of objects as representing the objects. I know that my cat doesn't. He is incapable of associating a photograph even of something interesting, like a fish, with the actual object. We can; we are sophisticated enough to possess such a higher mode of knowing about objects like chairs. From that to the methods of modern science I see no philosophically relevant discontinuity.

I think it does no good for scientists to pretend that we have a clear a priori idea of the scientific method. But still, we should try to say something about what it is that we think we are doing when we make progress toward truth in the course of our scientific work. There is one philosophical principle that I find of use here. It is, to paraphrase another author, "It don't mean a thing if it ain't got that zing." That is, there is a kind of zing—to use the best word I can think of—that is quite unmistakable when real scientific progress is being made.

Here is an example, a vicarious one, because I was not one of the active players at the time. Back in the mid-1950s, when I was a graduate student, particle physicists were beset with a tangle of problems associated with the properties of certain particles called K mesons. In 1956, two young theorists, Tsung Dao Lee and Chen Ning Yang, pointed out that the whole tangle could be resolved if you just took away one of the assumptions on which all previous analyses had been based, the assumption of what is technically called parity conservation—essentially just the symmetry between right and left. This symmetry principle states that the laws of nature will take the same form whether the x, y, and z axes of your coordinate system are arranged like the thumb, forefinger, and middle finger of your right hand or of your left hand. This had been taken for granted, by that time, for thirty years. It was regarded as self-evident. I, in my wisdom as a graduate student, thought it was absolutely absurd to challenge this principle, especially since it had been very successfully used in a wide variety of contexts in atomic and nuclear physics. Yet here were Lee and Yang proposing that this symmetry principle must be given up in order to understand the K mesons! They also proposed experiments that could verify their idea, and within months their idea was verified. Within less than a year the whole world of physics was convinced that Lee and Yang were right and that this thirty-year-old principle, which we had all taken for granted, was not universally applicable. In the following two or three years this discovery led to a great clarification in our understanding of the weak nuclear forces, going beyond the question of left-right symmetry. In showing that the question of whether or not a symmetry principle is true must be tested experimentally, Lee and Yang had reversed the tendency, beginning with Einstein, to take symmetries as given, nearly self-evident principles. The sense of excitement, of breakthrough, of accomplishment—in short, the "zing"—was evident to everyone.

Another example is closer to my own work. The 1960s had seen the development of a unified theory of weak nuclear forces and electromagnetism. It was not clear that this theory was mathematically consistent, although Abdus Salam and I argued that it was. Then, in 1971, a previously unknown graduate student, Gerard 't Hooft at the University of Utrecht, showed that theories of this type are, in fact, mathematically consistent. Immediately the world of theorists began to take this theory seriously and write many papers about it. It had not yet, however, become part of the scientific consensus. This change in the consensus took longer than the one started by Lee and Yang. In 1973, two years after 't Hooft's first work, and six years after my own earlier work, experimental evidence began to emerge showing that the theory was valid. Even so, although the theory was now widely held to be right, there remained some healthy skepticism, which was reinforced in 1976 when some other experiments pointed in the other direction. Finally, in 1978, experiments done at the Stanford Linear Accelerator Center decisively supported the unified theory of weak and electromagnetic forces, and it was from then on generally regarded as correct. From the very beginning to the end, the process of general acceptance had taken about eleven years, of which five were a period of intense experimental effort.

Experiment always has something to do with the fashioning of a scientific consensus, but in ways that can be quite complicated. In this case, the theorists were pretty well convinced of the general idea of this sort of theory after 't Hooft's work in 1971, before there was the slightest new experimental evidence. The rest of the physics community became convinced over a longer period of time, as the experimental evidence became unmistakable. But by the end of the 1970s there was a nearly universal consensus that this theory was right.

This story illustrates a few points. First, the interaction between theory and experiment is complicated. It is not that theories come first and then experimentalists confirm them, or that experimentalists make discoveries that are then explained by theorists. Theory and experiment often go on at the same time, strongly influencing each other.

Another point, ignored almost always by journalists and often by historians of science, is that theories usually exist on two levels. On the one hand, there are general ideas, which are not specific theories, but frameworks for specific theories. One example of such a general idea is the theory of evolution by natural selection, which leaves open the question of the mechanism of heredity. In the case of the unified theory of weak and electromagnetic forces, the underlying general idea is that the apparent differences between these forces arise from a phenomenon known as "spontaneously broken symmetry," that the equations of the theory have a symmetry between these forces, which is lost in the solution of these equations—the actual particles and forces we observe. These general ideas are very hard to test because by themselves they do not lead to specific predictions. This has sadly led Karl Popper to conclude that because such general ideas can't be falsified, they can't be regarded as truly scientific.

Then there are the specific, concrete realizations of such ideas. These are the theories that can be tested by experiment, and can be falsified. As it happened, with the unified weak and electromagnetic theory the symmetry pattern that had been originally suggested as a specific realization of the general idea of broken symmetry (aspects of which can be found also in work of Sheldon Glashow and Salam and John Ward, which did not incorporate this idea) turned out to be the right one. During the period of the 1970s, theorists were mostly convinced of the general idea, but not of this specific realization of the idea. Of course, the experimentalists had to prove that some specific theory was right before any of this could become part of the scientific consensus.

Also, when I say that the physics community became universally convinced of something, I am speaking loosely—this is never entirely true. If you had a lawsuit that hinged on the validity of the unified weak and electromagnetic theory, you could probably find an expert witness who is a Ph.D. physicist with a good academic position who would testify that he didn't believe in the theory. There are always some people on the fringes of science, not quite crackpots, often people with good credentials, who don't believe the consensus. This makes it harder for an outsider to be sure that the consensus has occurred, but it does not change the fact of the consensus. The consensus for Lee and Yang's idea about the symmetry between right and left and the consensus for the unified theory of weak and electromagnetic interactions were unmistakable when they happened.

I would also like to point out that, at least within the area of physics, which is what I mostly know about, and within this century, whenever this consensus has been achieved, it has never been simply wrong. To be sure, sometimes the truth turns out to be more complicated than what had been thought. For example, before 1956 there had been a consensus that there is an exact symmetry between right and left, and then we learned that the symmetry is not exact. But it is a good approximation in certain important contexts. The thirty years of theoretical physics research that relied on that symmetry to understand nuclear and atomic problems was not wrong; there were just small corrections that physicists didn't know about. No consensus in the physics community has ever been simply a mistake, in the way that in earlier centuries you might say, for example, that the theory of caloric or phlogiston was a mistake.

Now, all of this is of course a social phenomenon. The reaching of consensus takes place in a worldwide society of physicists. This fact has led to a second fallacy: that, because the process is a social one, the end product is in whole or in part a social construct.

Not even the social milieu of physics research is well described by postmodern commentators. It is far less oppressive and hegemonic than many would imagine. In many cases the great breakthroughs are made by youngsters like 't Hooft, of whom no one has heard before, while the famous graybeards who have senior positions in the leading universities often get left behind. Werner Heisenberg and (to a lesser extent) Paul Dirac were left behind by the physics community after 1945, as were Einstein and Louis de Broglie after 1925. Heisenberg and de Broglie rather discreditably tried to force their views on the physics communities in Germany and France. Einstein and Dirac, gentler souls, simply went their own ways. But even Heisenberg and de Broglie were not able to damage German or French physics for very long. The exact sciences show a remarkable measure of resilience and resistance to any kind of hegemonic influence, perhaps more than any other human enterprise.

The working philosophy of most scientists is that there is an objective reality and that, despite many social influences, the dominant influence in the history of science is the approach to that objective reality. It may seem that, in asserting the objective validity of what we are doing, scientists are simply trying to protect their own status. It is not easy to answer that criticism. I could say, "I am not a crook," but such arguments only go so far. Perhaps the best answer is "tu quoque." It seems to me that much of the comment on science by the social constructivists and postmodernists is motivated by the desire to enhance the status of the commentator—that he be seen not as a hanger-on or adjunct to science, but as an independent investigator, and perhaps as a superior investigator, by reason of his greater detachment. I think that this is especially true of those who follow the "Strong Program" in the sociology of science.

This motivation was close to the surface in a 1991 article in *ISIS* by Paul Forman. He described historians of science as preoccupied with their independence from the sciences. He called for a greater degree of independence, because this was important to their work as historians. So far, so good, but he also wanted historians to exercise an independent judgment not just as to how progress is made, which certainly is in their province, but also on whether progress is made. He gave no arguments that such judgments would have any kind of intellectual validity, except that this was the sort of thing that historians have to do as part of being historians.

I think we scientists need make no apologies. It seems to me that our science is a good model for intellectual activity. We believe in an objective truth that can be known, and at the same time we are always willing to reconsider, as we may be forced to, what we have previously accepted. This would not be a bad ideal for intellectual life of all sorts.

9

Night Thoughts
of a Quantum Physicist

The American Academy of Arts and Sciences was founded in
Boston in 1780 by John Adams. (It is said that Adams was uphold-
ing the good name of Boston in response to the formation of the
American Philosophical Society in Philadelphia.) From the time I
became a member of the American Academy in 1967 until I left the
Greater Cambridge area for Austin in 1982, the Academy gave me
a wonderful opportunity for meeting interesting people who were
not physicists. Academy members and spouses meet once a month
during the academic year for drinks, dinner, and a talk on some
learned subject. When I was first a member these meetings were
held at Brandegee House, a luxurious imitation Renaissance pa-
lazzo built in the nineteenth century by a Boston merchant prince.
It was a great pleasure for me to speak there in 1979, on the occasion of the centennial of Einstein's birth. The venue of these meetings later shifted to the Academy's handsome but more austere
present house, built in Cambridge through the generosity of Edwin
Land. I was invited back to Cambridge to speak in the present
house of the Academy in February 1995; the essay below, published in the *Bulletin* of the Academy, is based on that talk.

On this occasion I said that in twenty years we had not made
any progress toward a fundamental understanding of nature
within the old paradigm of quantum field theory. We had been
looking for several decades at so-called string theories, which seem
to be our best hope for a more fundamental theory. But no string
theory had jelled into something that could be tested experimentally. Alas, I could say the same today, except that the figure of

twenty years is now twenty-five years. But there are now some new reasons for hope. There is now experimental evidence of neutrino masses that if confirmed would provide the first sign of something beyond our present Standard Model. And we now have theoretical evidence that all the different string theories explored in the 1980s and early 1990s are merely different approximations to a single theory. As for the European accelerator I mentioned in this talk, the Large Hadron Collider at CERN, work has continued, and it is now pretty definite that the LHC will go on line in 2005.

After the publication of this talk an error of mine was forcefully pointed out by a participant in what is called the Strong Program in the sociology of science, a program begun at the University of Edinburgh that emphasizes (and I would say overemphasizes) cultural influences on the work of science. I should not have said in my talk that adherents of this program argue that scientific theories are "nothing but" social constructions, for these social constructivists acknowledge that our theories are constrained in part by objective reality. In my essay "Physics and History" later in this collection I have tried to be more precise in describing what I object to in the Strong Program and in similar tendencies in the history and sociology of science.

I think many of you will have recognized the source of this talk's title: a lovely novel by Russell McCormmach called *Night Thoughts of a Classical Physicist*. It's about a fictitious physicist, Victor Jakob, who in 1918 looked back over his career, which had spanned the first two decades of the twentieth century, and described his sense of frustration and unease with what was happening to physics. Night thoughts, of course, are what you get when you wake up at three in the morning and can't figure out how you are going to go on with your life.

During the early years of the twentieth century, all sorts of things were changing in ways that disturbed an old-fashioned physicist like Victor Jakob. Many things were known experimentally—for example, atomic spectra filled books of data—but they weren't understood. Niels Bohr and Arnold Sommerfeld and others had invented mysterious rules that allowed you to calculate some of the wavelengths listed in those spectral tables, but nobody knew why the rules worked, and they didn't always work, so they didn't make much sense. And even worse, in the area that classical physicists thought they understood best—gravity and space and time, the area that Newton had so wonderfully opened up— changes were in the air. Space and time were now somehow tied to gravity. Gravity was not a force; it was a curvature in space and time. Victor Jakob was not happy with this and so had some rather bad night thoughts.

Since then there has been wonderful progress in physics (and in science generally, of course, but I know mostly about physics). In the 1920s the atoms that so puzzled physicists of Jakob's generation were explained. All those wavelengths that filled books of spectroscopic tables were explained—not in terms of ad hoc, mysterious rules that sometimes worked and sometimes didn't, but in terms of quantum mechanics, a clear, coherent, understandable framework for physics. Quantum theory—which Jakob might have been able to learn, had he lived until 1925—overturned classical ideas of determinism and provided an entirely new description of nature. The atom became understood entirely in terms of the principles of quantum mechanics and a few properties of electrons and of the atomic nuclei that form most of the mass of atoms.

In the years since the mid-1920s, we have come to understand what atomic nuclei are in terms of more fundamental particles. By the 1970s all the properties of matter were, in principle, explicable in terms of a relatively simple, coherent theory—a mathematically consistent theory called the Standard Model.

Nevertheless, standing here near the end of the twentieth century, I must tell you that even with the benefit of all that glorious progress, we physicists of this moment have our night thoughts, too. But they are rather different from—in fact, in some respects they are opposite to—those of Victor Jakob.

You are probably all familiar with the story that Alexander the

Great wept because he had discovered that there were no more worlds for him to conquer. Experts tell me that this story is apocryphal; it's not in any of the early accounts of Alexander's life. Even without these expert opinions, I had always felt that that there was something fishy about this story: Alexander couldn't possibly have thought that he had conquered the whole world. He knew about the Ganges River in India, for example—and he knew he hadn't reached it. He certainly knew, as did every Greek, about the greater Greece in Sicily—and he hadn't conquered Sicily. So what does that story mean? The story does make sense if you tell it with the emphasis on the right word; Alexander wept because there were no more worlds for *him* to conquer.

Today we particle physicists are in Alexander's position. We have already conquered a great many worlds. We understand atoms and atomic nuclei and the particles that make up atomic nuclei and the particles that make up those particles. All the mysteries of the nature of matter and force and energy have been crammed into the framework of the Standard Model. But the Standard Model is clearly not the final answer, and we seem to be unable to take the next step. I could give you a talk now about the Standard Model, and it would not differ materially from a talk I might have given on that subject twenty years ago. In the past fifteen or twenty years, there's really been no essential progress in physics, in the sense of discovering more fundamental theories which are then confirmed by confrontation with experimental data.

Briefly, the Standard Model is a field theory. I think you all have some idea of what fields are. For instance, you've all seen pictures of iron filings on a magnet, and you know that a magnet creates a condition of space around it called a magnetic field. You can feel that field if you hold a piece of iron in your hand and extend it toward a magnet. Similarly, the earth creates a condition of space around it called a gravitational field. When quantum mechanics is applied to such fields, we learn that the energy and momentum and other properties of the fields are not spread out uniformly;

they come in little bundles, or quanta. Those quanta are what we recognize as particles.

In quantum field theory, the fundamental ingredients of nature are fields, and all the particles that are the constituents of atoms, or the constituents of the constituents of the constituents of atoms—particles like electrons and quarks—are bundles of field energy. In the Standard Model, there are a few dozen types of field, and all the particles are just epiphenomena—secondary manifestations of those fields.

Twenty years ago, if you had asked me about the shape of the next fundamental theory, I would have said it would be a better field theory that somehow tied things up more elegantly mathematically and enabled us to understand better why things are what they are. That theory has not materialized. Meanwhile, we know that the Standard Model is not the final answer, because of its obvious imperfections—and those imperfections, I have to say, are aesthetic. The Standard Model accounts successfully for everything we know that has been discovered in accelerator laboratories designed to test it, but it's clearly imperfect. As I said earlier, it involves a large assortment of fields—a few dozen—but we don't know why we have those fields and not other fields. Furthermore, about eighteen numbers, representing certain constants of nature (for example, the mass of the electron) appear in the equations of the Standard Model, but in this theory they can't be explained; they simply have to be learned from experiment. It's a little bit much to say that the fundamental theory of nature has eighteen free parameters that cannot be explained.

Also, the Standard Model leaves out one rather important ingredient: gravitation. Gravity just doesn't fit into the Standard Model. This is one of the reasons we think that the structures we describe in the Standard Model are not going to be the structures of the next fundamental theory in physics.

In the past twenty years, we physicists have felt, to some extent, as if the earth has moved under our feet. We thought that in quantum field theory we had a firm foundation for future developments in physics, and then we began to realize that we don't. We are gradually awakening to the fact that any theory that satisfies certain fundamental axioms—the axioms of relativity, the axioms of quantum mechanics, and a few others that seem inescapable—looks, at sufficiently large scales of distance, like a quantum field theory. In other words, the success of our quantum field theory doesn't prove that it's really a fundamental theory, because any theory, when studied at sufficiently large distance scales—for example, those found inside atomic nuclei, which to today's particle physicists seem pretty big—looks like a quantum field theory.

My current work deals with the classic problem of trying to understand nuclear forces. In doing this work, I employ a field theory in which one of the fields is called the pi meson field. The pi meson is a particle that was discovered in cosmic rays in 1947. It's believed to play an important role in producing forces inside the nucleus. We have known for some years that the pi meson is not a fundamental particle; it's a composite of quarks and antiquarks. But in dealing with nuclear forces, it makes sense to work with a field theory of pi mesons as our starting point, because *any* theory, whether it's about quarks and antiquarks or whatever, when looked at in terms of the relatively large scale of distances that you find inside atomic nuclei, looks like a field theory, and given certain invariance principles, it must look like a field theory of pi mesons. So we do not interpret the success of this field theory as telling us that the pi meson field would necessarily be an ingredient in a truly fundamental theory. And, by extension, we cannot interpret the success of the Standard Model as indicating that the fields of electrons and quarks and so on are fundamental entities.

This sort of revolution in what we take as fundamental has happened before in the history of physics. When Einstein's General Theory of Relativity replaced Newton's theory of gravity, it didn't replace it by finding small corrections to the inverse square law. It replaced it by eliminating the fundamental conception of Newton's theory: that gravity was a force exerted on one body by another. In General Relativity, you don't talk about force; you talk about a curvature of space and time. The effect of the replacement of Newtonian mechanics by General Relativity on predictions about the solar system was to introduce corrections of less than one part in a million, but Einstein's theory revolutionized our way of describing nature.

Now we need another revolution in our way of describing nature. In twenty years we haven't made any progress in the old paradigm of quantum field theory. In particular, we haven't figured out a way of describing gravitation in quantum field theory. If you just go ahead and try to describe gravitation by the same sort of quantum field theory that is used for other forces, then when you use this theory to answer perfectly sensible questions about gravitational processes, the answer is usually infinity—which is not a sensible answer. And for that reason, we have largely given up trying. Clearly, if we are going to bring gravity into our theories, we need some conceptual framework other than quantum field theory. Where are the deeper structures of this theory to be found? The things within the particles within the particles within the particles inside the atom? How much more finely do we have to explore nature?

As everyone knows, Democritus and his teacher Leucippus speculated about atoms. After more than two millennia, it was discovered that atoms do exist. Although Democritus didn't know it, there was a factor of about ten billion between the scale of lengths with which he was familiar and the scale at which atoms were to be found. Bridging that gap took more than two millennia. Then from atoms to the atomic nucleus, there was another factor of a hundred thousand: the atomic nucleus is about a hundred-thousandth as big as the atom itself. The lengths probed in today's elementary particle physics laboratories are about a hundred times smaller than an atomic nucleus. At that scale, everything is very well described by quantum field theory, by the Standard Model.

How much deeper do we have to go to find the deeper structures that underlie the standard model? We don't know for sure. They may be just around the corner. But there is good reason to believe that these structures are much smaller than the smallest structures we have studied so far—not by a factor of a hundred thousand or ten billion, but by a factor of a million billion (10^{15}). In other words, whatever they are, these structures are much smaller compared with the size of an atomic nucleus than an atomic nucleus is compared with a person. This presents a problem: we cannot directly study such structures experimentally. So, in our efforts to make progress, we pursue two other approaches.

One approach, pursued by a younger generation of theoretical physicists, is known as string theory. So far, string theory has been mostly driven by a sense of mathematical beauty and by a search for consistency. It has wonderful features, including a natural role for gravity—in fact, it requires the existence of gravity. But it has made no new predictions that have been experimentally verified. It is the first candidate we have for a final theory, but we have been looking at string theories for several decades now, and they haven't yet jelled into one specific theory that can be tested experimentally. String theory is mathematically very difficult. I think it's a sign of the intrinsic health of physics that, even though most physicists of the older generation don't learn string theory and can't read papers about it, young string theorists continue to get tenure at leading American universities. Our field is not (as some imagine it to be) hegemonic, dominated by an old guard of fuddy-duddies. It is very much alive to new possibilities.

The other tack is experimental. We want the inspiration of new experimental results that can put us on the right track toward the fundamental string theory—or, if it's not a string theory, toward whatever it is. These experiments would not be able to explore structures a million billion times smaller than those studied today; rather, they would seek to complete the Standard Model, which still has some uncertain aspects having to do with how particles get their masses. Understanding those aspects of the Standard Model might give us just the kick in the pants we need to start us moving again toward the next deeper theory.

We were hoping that the Superconducting Super Collider, which would have been built in Ellis County, Texas, would help provide that kick in the pants, but its construction has been canceled. We hope that the Europeans will continue (and they now seem to be definitely deciding that they will continue) with building another accelerator, which will have about a third of the power that the Super Collider would have had and will be finished five years later than the Super Collider would have been—sometime around the year 2004.

I have been giving you what a sociologist of science would call an internalist view. I have been describing what has been happening entirely by following the logical imperative of our theories and of our experimental capabilities. The need for money for increasingly expensive experiments, of course, puts us very much under the influence of external factors.

In looking at the outside world, today's scientists—like monks in the time of Henry VIII, I guess—have begun to suspect that they are not universally loved. This, of course, is an old story. You can go back to Jonathan Swift and William Blake and Walt Whitman and find expressions of strong antiscientific sentiments. But what we see now are active elements within our universities that are hostile to the pretensions and the goals of science in general or particle physics in particular—and I want to say a little bit about that, although it is not an area of my expertise, and Gerald Holton has written about this in a much more authoritative way.

The debate about the value of particle physics itself has taken place largely within the scientific community. We particle physicists don't claim that ours is the only work worth supporting. I suppose we might if we thought we could get away with it, but in fact we don't believe that such a claim would be true. We do believe that our work has a unique sort of value, because we work at the most fundamental level in all of science. (Incidentally, *particle physicist* is a terrible term because we are not really that interested in particles; we're interested in principles. We use particles as tools to get at the underlying scientific principles. But for want of a better label, call us particle physicists).

If you ask any question about why things are the way they are, part of the answer, if you are talking about biology or astronomy, is probably going to be based on historical accidents, like the details of how the solar system formed. If we leave historical accidents aside, however, the answer will concern general principles. But there are no freestanding principles of biology or chemistry or economics. If you discover a law of chemistry or a law of economics, you then must ask why it is true. The explanation always takes the form of a reduction to a deeper theory. Deeper not in the sense of being mathematically more profound or useful—often, it isn't— but in the sense of being closer to the origin of our explanations.

In the same way that much of mineralogy and physiology is explained in terms of chemistry, and chemistry is explained in terms of physics, the physics of ordinary matter is explained in terms of the Standard Model. That doesn't necessarily mean that it is actually explained, however. Very often, the explanation in terms of the Standard Model is too complicated for us to be able to work it all out. But that's where you would finally look. You would not look for a freestanding, autonomous law of superconductivity or economic behavior. You would try to explain each of those in a reductionist way.

I get into terrible arguments about this view, and I hope I have expressed it in a sufficiently provocative way to stimulate some argument here. This pretension of particle physics has enraged other scientists—in particular, other physicists. You'll find endless nasty letters back and forth about it in the letters column in *Physics Today*.

Then there are attacks on the pretensions not just of particle physics but of science in general. Here again, if you don't mind repeated bifurcations, I would say that there are two broad strands. One is social constructivism. The social constructivists, including some of the historians of science and many of the sociologists of science, recognize that we reach a consensus about scientific discoveries through a social process. The works of Peter Galison and Sam Schweber that explore the history of developments in particle physics and field theory illustrate this sort of social negotiation about which competing theory or experiment is correct. This is valuable work to which no one could raise objections. By social constructivism I mean the more controversial view that what scientists do is largely conditioned by the social interactions of scientists with each other and with the larger society in which they live. In other words, while scientific discoveries may in a sense be about something objective, they wouldn't be what they are if it weren't for the Zeitgeist. A famous article by Paul Forman describes how the disillusionment of post–World War I Germany was the necessary framework for the development of quantum mechanics. I don't find this view illogical or obviously absurd, but I do think that it is greatly exaggerated. My own experience with science is that it is mostly directed by internal factors; the pull of reality is what makes us go the way we go. Of course, society has to offer the opportunity for progress, without which we would do nothing. But once given that opportunity, the direction we take is determined by external reality.

There is also a really radical social constructivist view—associated, for example, with the Strong Program initiated at the University of Edinburgh. According to this view, scientific theories are nothing but social constructions—which seems to me absurd. By the way, it's very difficult to sort out various versions of social constructivism, because everyone who works in the sociology or history of science seems to try very hard to distance himself or herself from everyone else. They each describe infinite gradations of belief so as not to seem to take the same position as someone else. I recall the observation of Stan Ulam, at a conference on arms control, that "madness is the ability to make fine distinctions among different kinds of nonsense."

Then there are the postmoderns. Whereas the social constructivists are serious people who I think are wrongheaded but who do a lot of good work in following the history of individual scientific developments—in fact, some of their articles are really quite illuminating—there are postmoderns who not only doubt the objectivity of science but dislike objectivity, who would welcome something warmer and fuzzier than modern science. These postmoderns are intellectual descendants of the heretic whom Thomas Aquinas most detested—Siger of Brabant. Siger argued that although immortality of the individual soul is not philosophically allowed and therefore, from a philosophical perspective, is untrue, it is theologically correct. Thus, it is both true and not true, depending on your mode of thinking. This seems to be the kind of reasoning that postmoderns would welcome.

By the way, in this respect my friend Thomas Kuhn has a lot to answer for. He distances himself from the postmoderns and the social constructivists, but he is endlessly quoted by them. He distances himself in saying that there is a place for evidence and reason in the scientific process—good to hear—but he attacks the idea that we are moving toward objective truth. As far as I can tell from one of his recent articles, his reason for rejecting the idea that science moves toward objective truth is that he and other philosophers have not succeeded in defining truth—and he cannot say what truth would be. This seems a bit like saying that because farmers cannot define cows or the difference between cows and, say, buffaloes, one should doubt the objective existence of cows. I would argue that it's not the job of farmers to define cows; that's the job of zoologists. Likewise, it's not the job of physicists or other scientists to define truth; that's the job of philosophers. If they haven't done that job, too bad for them. But just as the farmer generally knows cows when he sees them, we scientists usually know truth when we see it.

Another strand of antiscientific sentiment is the feminist criticism of science. There is no question that women have often been excluded from scientific work and from scientific organizations—even from this Academy. But according to one version of the feminist criticism of science, modern science is intrinsically masculine, particularly because of its insistence on objective truth and its insistence that some scientific theories are simply wrong. Proponents of this feminist view maintain that this androcentric, Western-oriented, reductionist science should be replaced with something that is feminist, Eastern-oriented, and holistic. Someone who took that position seriously might consider it a good reason for keeping women out of science—but I don't take it seriously, and I am very glad to observe that there is no perceptible difference in the way that men and women physicists do physics.

To put all this in perspective, I have to say that these arguments about the objectivity of science have not really gone outside the academy. For example, they haven't been important in Congress. In trying to get votes for the Superconducting Super Collider, I was very much involved in lobbying members of Congress, testifying to them, bothering them, and I never heard any of them talk about postmodernism or social constructivism. You have to be very learned to be that wrong.

Recently, I spoke at a meeting of an organization called the National Association of Scholars, which has been formed to oppose these and other antiscience trends. I remarked that I didn't think antiscientific sentiment was really much of a problem affecting support of science in Congress, and the audience got angry with me for not seeming sufficiently scared, until to make peace I had to say that I was scared too.

There is one development that really would scare me. If the articulate and influential antiscientific intelligentsia inside the university allies itself with the enormous force of religious belief outside the university, we will really have something to be scared about.

But I am not really that scared or that disillusioned. Bryan Appleyard, a British journalist, recently wrote *Understanding the Present: Science and the Soul of Modern Man*—a book that is quite intelligent and eloquent but also hostile to science. Part of his complaint about science is that it has diminished religious zeal and fostered liberal democracy. As a scientist, that is the sort of thing to which I am happy to plead guilty. Particle physicists, other physicists, and other scientists have been saying for some time that the product of our work is not merely arcane theories that we understand and others don't. It's not merely new devices, new medicines, new weapons. The product of our work is a world view that has led to the end of burning heretics and, if Appleyard is right, to the fostering of liberal democracy—or at least to an understanding that we are not living in a world with a nymph in every brook and a dryad in every tree. I feel that this, above all, is the thing about which we scientists can be most proud.

10

Reductionism Redux

This essay started as a response to a 1995 article in the *New York Review of Books* by the physicist Freeman Dyson. I have been an admirer of Dyson for decades, ever since as a graduate student I learned the quantum theory of fields by reading Dyson's famous 1949 papers in *The Physical Review.* In addition to being a distinguished theoretical physicist he is an eloquent writer, the author of *Disturbing the Universe* and other books. We are on good terms personally, and Dyson had nice things to say in his article about my work, but he is also someone whose opinions on all sorts of matters, from religion to the building of accelerators, tend to be different from my own.

In his 1995 article Dyson attacked the influence of reductionism in scientific research. Unlike most others who take this tack, Dyson at least gave a fair description of the aim of reductionism in physics: "to reduce the world of physical phenomena to a finite set of fundamental equations." But he challenged the value of this sort of physics, and for illustration he pointed to a failure in the scientific career of J. Robert Oppenheimer.

I didn't find this welcome. My own work is very much in the style that Dyson calls reductionist, and if I didn't think that this was a good thing I would be doing something else. I didn't even agree with what Dyson said about Oppenheimer.

So I wrote to Robert Silvers, the editor of *The New York Review,* introducing myself and volunteering to write an essay defending reductionism from attacks by Dyson and others. Silvers didn't take to this idea, but he suggested that instead I might write an essay review of the book[1] in which Dyson's article had originally appeared. This book had grown out of a small meeting at Jesus College, Cambridge, that in 1992 had brought together philosophers and scientists to discuss reductionism.

I didn't think much of my ability as a book reviewer. I had written only one book review before, a rather stodgy piece for the Sunday book section of *The New York Times*, and I hadn't been asked to repeat the performance. But I had a special admiration for *The New York Review*, especially for the distinction of its authors, the breadth of its coverage, and the luxurious length of its articles, so I decided to have a go at this review.

This began my connection with *The New York Review*, a collaboration that has lasted to the present and has been responsible for five of the essays in this collection. One of the best parts of this has been the experience of working with Robert Silvers. Many editors do little beyond routine copy editing, while others try to rewrite your work, putting your words in their words. (The editors of *Scientific American* used to be particularly active this way.) In contrast, Silvers gives every article a close reading, and makes dozens of wise detailed suggestions, written in the margin in his execrable handwriting, but he never tries to replace your ideas or style with his own. How he does this with all the articles published in *The New York Review* I can't imagine.

It used to be traditional for college courses on the history of philosophy to begin around 600 B.C.E. with Thales of Miletus. According to later writers, Thales taught that everything is made of water. Learning about Thales, undergraduates had the healthy experience of starting their study of philosophy with a doctrine that they knew to be false.

Though wrong, Thales and his pre-Socratic successors were not

1. *Nature's Imagination: The Frontiers of Scientific Vision*, ed. John Cornwell, with an introduction by Freeman Dyson (Oxford: Oxford University Press, 1995).

just being silly. They had somehow come upon the idea that it might he possible to explain a great many complicated things on the basis of some simple and universal principle—everything is made of water, or everything is made of atoms, or everything is in flux, or nothing ever changes, or whatever. Not much progress could he made with such purely qualitative ideas. Over two thousand years later Isaac Newton at last proposed mathematical laws of motion and gravitation, with which he could explain the motions of the planets, tides, and falling apples. Then in the *Opticks,* he predicted that light and chemistry would someday be understood "by the same kind *of* reasoning as for mechanical principles," applied to "the smallest particles of nature."

By the end of the nineteenth century physicists and chemists had succeeded in explaining much of what was known about chemistry and heat, on the basis of certain assumed properties of some ninety types of atoms—hydrogen atoms, carbon atoms, iron atoms, and so on. In the 1920s physicists began to be able to explain the properties of atoms and other things like radioactivity and light, using a new universal theory known as quantum mechanics. The fundamental entities to which this theory was applied were no longer the atoms themselves but particles even more elementary than atoms—electrons, protons, and a few others—together with the fields of force that surround them, like the familiar fields that surround magnets or electric charges.

By the mid-1970s it had become clear that the properties of these particles and all other known particles could be understood as mathematical consequences of a fairly simple quantum theory, known as the Standard Model. The fundamental equations of the Standard Model deal not with particles and fields, but with fields of force alone; particles are just bundles of field energy. From Newton's time to our own we have seen a steady expansion of the range of phenomena that we know how to explain, and a steady improvement in the simplicity and universality of the theories used in these explanations.

Science in this style is properly called reductionist. In a recent article in these pages[2] Freeman Dyson described reductionism in physics as the effort "to reduce the world of physical phenomena to a finite set of fundamental equations." I might quibble over whether it is equations or principles that are being sought, but it seems to me that in this description Dyson has caught the essence of reductionism pretty well. He also cited the work of Schroedinger and Dirac on quantum mechanics in 1925 and 1927 as "triumphs of reductionism. Bewildering complexities of chemistry and physics were reduced to two lines of algebraic symbols."

You might have thought that these illustrious precedents would inspire a general feeling of enthusiasm about the reductionist style of scientific research. Far from it. Many science kibitzers and some scientists today speak of reductionism with a sneer, like postmodernists talking about modernism or historians about Whig historiography. In 1992 John Cornwell, the director of a project on the sociology of science at Jesus College, Cambridge, convened a group of well-known scientists and philosophers to meet there to discuss reductionism. It was at this symposium that Dyson gave the talk on which his eloquent *New York Review* article was based. The collected papers of this symposium appear in *Nature's Imagination*,[3] which contains articles with titles such as "Must Mathematical Physics Be Reductionist?" (Roger Penrose), "Reductive Megalomania" (Mary Midgley), and "Memory and the Individual Soul: Against Silly Reductionism" (Gerald Edelman). A review of this book by the mathematician John Casti, in *Nature,* calls these authors the "good guys in the white hats" as opposed to the unreconstructed reductionists at the meeting like the chemist Peter Atkins and the astronomer John Barrow.

Casti is a fellow of the Santa Fe Institute, a haven for nonreductionist science. Dyson himself remarks that he has a "low opin-

2. Freeman Dyson, "The Scientist as Rebel," *The New York Review of Books,* May 25, 1995, pp. 31–33.

3. This book contains interesting articles on the foundations of mathematics and on other subjects, which I will not discuss here because I want to concentrate on reductionism in the natural sciences.

ion" of reductionism. (Coming from Dyson, this really hurts. He played a major role in the development of quantum field theory, which has been the basis of the reduction of all of elementary particle physics to the Standard Model.) What has gone wrong? How has one of the great themes in intellectual history become so disreputable?

One of the problems is a confusion about what reductionism is. We ought first of all to distinguish between what (to borrow the language of criminal law) I like to call grand and petty reductionism. Grand reductionism is what I have been talking about so far—the view that all of nature is the way it is (with certain qualifications about initial conditions and historical accidents) because of simple universal laws, to which all other scientific laws may in some sense be reduced. Petty reductionism is the much less interesting doctrine that things behave the way they do because of the properties of their constituents: for instance, a diamond is hard because the carbon atoms of which it is composed can fit together neatly. Grand and petty reductionism[4] are often confused because much of the reductive progress in science has been in answering questions about what things are made of, but the one is very different from the other.

Petty reductionism is not worth a fierce defense. Sometimes things can be explained by studying their constituents—sometimes not. When Einstein explained Newton's theories of motion and gravitation, he was not committing petty reductionism. His explanation was not based on some theory about the constituents of anything, but rather on a new physical principle, the General Principle of Relativity, which is embodied in his theory of curved spacetime. In fact, petty reductionism in physics has probably run its course. Just as it doesn't make sense to talk about the hardness or temperature or intelligence of individual "elementary" parti-

4. Grand and petty reductionism correspond more or less to what the evolutionary biologist Ernst Mayr has called "theory reductionism" and "explanatory reductionism" in his article "The Limits of Reductionism," *Nature* 331 (1987), p. 475.

cles, it is also not possible to give a precise meaning to statements about particles being composed of other particles. We do speak loosely of a proton as being composed of three quarks, but if you look very closely at a quark you will find it surrounded with a cloud of quarks and antiquarks and other particles, occasionally bound into protons; so at least for a brief moment we could say that the quark contains protons. It is grand reductionism rather than petty reductionism that continues to be worth arguing about.

Then there is another distinction, one that almost no one mentions, between reductionism as a program for scientific research and reductionism as a view of nature. For instance, the reductionist view emphasizes that the weather behaves the way it does because of the general principles of aerodynamics, radiation flow, and so on (as well as historical accidents like the size and orbit of the earth), but in order to predict the weather tomorrow it may be more useful to think about cold fronts or thunderstorms. Reductionism may or may not be a good guide for a program of weather forecasting, but it provides the necessary insight that there are no autonomous laws of weather that are logically independent of the principles of physics. Whether or not it helps the meteorologist to keep it in mind, cold fronts are the way they are because of the properties of air and water vapor and so on, which in turn are the way they are because of the principles of chemistry and physics. We don't know the final laws of nature, but we know that they are not expressed in terms of cold fronts or thunderstorms.

One can illustrate the reductionist world view by imagining all the principles of science as being dots on a huge chart, with arrows flowing into each principle from all the other principles by which it is explained. The lesson of history is that these arrows do not form separate disconnected clumps, representing sciences that are logically independent, and they do not wander aimlessly. Rather, they are all connected, and if followed backward they all seem to branch outward from a common source, an ultimate law of nature that Dyson calls "a finite set of fundamental equations." We say that one concept is at a higher level or a deeper level than another if it is governed by principles that are further from or closer to this common source. Thus the reductionist regards the general theories governing air and water and radiation as being at a deeper level than theories about cold fronts or thunderstorms, not in the sense that they are more useful, but only in the sense that the latter can in principle be understood as mathematical consequences of the former. The reductionist program of physics is the search for the common source of all explanations.

As far as I can tell, Dyson's objections are entirely directed at reductionism as a research program rather than as a world view. He regrets that Einstein and (in later life) Oppenheimer were not interested in something as exciting as black holes, and blames this on their belief that "the only problem worthy of the attention of a serious theoretical physicist was the discovery of the fundamental equations of physics." This is pretty mild criticism. Dyson does not question the value of the discovery of fundamental equations (how could he?) but only tells us that there are other things in physics to think about, like black holes. This is like a prohibitionist who is against gin because, good as it is, it distracts people from orange juice. And I am not sure that Dyson is even entirely right about Einstein and Oppenheimer as examples of the dangers of the appeal of reductionism.

I recall as a Princeton graduate student going to seminars at the Institute for Advanced Study, where Dyson was a professor and Oppenheimer the director. I always sat in back and kept quiet, while Oppenheimer always sat in front and carried on a detailed technical dialogue with the speaker, whatever the topic might be. He certainly seemed interested in everything that was going on in physics, not just things at the reductive forefront. In fact, even in the 1920s and 1930s, when he was doing his best research, Oppenheimer's work had much less to do with finding fundamental equations than with calculating the consequences of existing theories. By the time I met him, Oppenheimer's own research had pretty well ended, and I can believe that he explained this even to himself in the way that is cited by Dyson; but I suspect that the truth is that he had just become too famous and too busy to have time for research.

Einstein is another story. He had never immersed himself the way Oppenheimer did in the physics research of others. The physicist-historian Gerald Holton showed some years ago that Einstein was not significantly influenced by the specific experimental result of Albert Michelson and Edward Morley that is often described as the crucial clue that led to Special Relativity. I think that Einstein objected to black holes not because he found them uninteresting but rather for precisely the reason that I and many others find them interesting; they suggested an incompleteness in his beloved General Theory of Relativity. Physics in the reductive style had served Einstein magnificently well until the 1920s, and he was not so much wrong in trying to continue in this vein as he was in assuming that the appropriate subjects for fundamental research would continue to be what they had been in his youth: gravitation and electromagnetism. He became narrow, endlessly pursuing the false goal of unifying gravitation and electromagnetism, and cut off from the exciting work on cosmic rays and elementary particles and quantum field theory that eventually led to the unification of the Standard Model. His real mistake is one we all risk: he became old.

Much of the criticism of reductionism is really only criticism of reductionism as a program for research. A good example is an argument by the moral philosopher Mary Midgley. In her article in the collection based on the Jesus College symposium, she asks, "What, for instance, about a factual statement like 'George was allowed home from prison at last on Sunday?' How will the language of physics convey the meaning of 'Sunday'? or 'home' or 'allowed' or 'prison'? or 'at last'? or even 'George'?" This criticism would strike home if there were physicists who were trying to use physics for such a purpose, but I don't know of any.

It is not just that (as emphasized in this symposium by Atkins and the philosophers Paul and Patricia Churchland) prisons and people and thunderstorms are too complicated for us to be able to predict their behavior by following the motions of the elementary particles of which they are composed. It is also a matter of what interests us. The buzzword here is "emergence." As we deal with more and more complicated systems, we see phenomena emerge from them that are much more interesting than a mountain of computer printout describing the motion of each particle in the system ever could be. Mind is a phenomenon that emerges from the biology of complicated animals, just as life is a phenomenon that emerges from the chemistry of complicated molecules. We are interested in whether George is happy to be out of jail in a way that is different from our interest in his nerve cells, and we are interested in his nerve cells in a way that is different from our interest in the electrons and protons and neutrons of which they are composed. But phenomena like mind and life do *emerge*. The rules they obey are not independent truths, but follow from scientific principles at a deeper level; apart from historical accidents that by definition cannot be explained, the nervous systems of George and his friends have evolved to what they are *entirely* because of the principles of macroscopic physics and chemistry, which in turn are what they are *entirely* because of the principles of the Standard Model of elementary particles.

It is not so much that the reductionist world view helps us to understand George himself as that it rules out other sorts of understanding. Every field of science operates by formulating and testing generalizations that are sometimes dignified by being called principles or laws. The library of the University of Texas has thirty-five books called *Principles of Chemistry* and eighteen books with the title *Principles of Psychology*. But there are no principles of chemistry that simply stand on their own, without needing to he explained reductively from the properties of electrons and atomic nuclei, and in the same way there are no principles of psychology that are freestanding, in the sense that they do not need ultimately to be understood through the study of the human brain, which in turn must in the end be understood on the basis of physics and chemistry. Henri Bergson and Darth Vader notwithstanding, there is no life force. *This* is the invaluable negative perspective that is provided by reductionism.

I suppose Midgley might retort that she doesn't know any anti-reductionist philosophers who think that there are freestanding principles of psychology. Maybe not, though many of our fellow citizens still think that George behaves the way he does because he has a soul that is governed by laws quite unrelated to those that govern particles or thunderstorms. But let that pass. In fact, I suspect that Midgley shares the world view of grand reductionism, but holds it "not honesty to have it thus set down."

At any rate, Midgley has to reach in some peculiar directions in her search for horrible examples of reductionism. One of her targets is B. F. Skinner, the late arch-behaviorist and master pigeon trainer. I share her dislike of Skinner's refusal to deal with consciousness in his work. But why does she quote him in a critique of reductionism? I am not aware that Skinner thought very much about sciences like evolutionary biology or neurology, which might provide reductive explanations for principles of psychology. I always thought that Skinner's problem was not reductionism, but positivism, the doctrine that science should concern itself only with what can be directly observed, like behavior. Positivism generally leads *away* from reductionism; for instance, at the beginning of this century it led the influential Viennese physicist-philosopher Ernst Mach to reject the idea of atoms, because they could not be directly observed.

Perhaps I do know why Midgley chose Skinner as a reductionist villain. Skinner excluded consciousness from his view of the mind, and consciousness poses the greatest challenge to reductionism. It is difficult to see how the ordinary methods of science can be applied to consciousness, because it is the one thing we know about directly, not through the senses.

Peter Atkins gave a splendid in-your-face reductionist polemic at Jesus College that I thoroughly enjoyed reading. "Scientists, with their implicit trust in reductionism, are privileged to be at the summit of knowledge, and to see further into truth than any of their contemporaries." Give 'em hell, Peter! But it seems to me that Atkins is not sufficiently sensitive to the problems surrounding consciousness. I don't see how anyone but George will ever know how it feels to be George. On the other hand, I can readily believe that at least in principle we will one day be able to explain all of George's behavior reductively, including what he says about how he feels, and that consciousness will be one of the emergent higher-level concepts appearing in this explanation.

In their articles in the symposium report the neuroscientists Gerald Edelman and Oliver Sacks make what I think is too much of the supposed antireductionist implications of new ideas about the brain. In his article, written with Giulio Tononi, Edelman describes his "Theory of Neuronal Group Selection," according to which the brain operates not according to a preset program, but rather according to one that evolves through a sort of natural selection during the life of the organism. He then argues in another article in this collection that

> the kind of reductionism that doomed the thinkers of the Enlightenment is confuted by evidence that has emerged both from modern neuroscience and from modern physics. I have argued that a person is not explainable in molecular, field-theoretical, or physiological terms alone. To reduce a theory of an individual's behavior to a theory of molecular interactions is simply silly. . . . Even given the success of reductionism in physics, chemistry, and molecular biology, it nonetheless becomes silly reductionism when it is applied exclusively to the matter of the mind.

Edelman is a very distinguished scientist, and his "neural Darwinism" may well be a great advance in the theory of the mind; but when he discusses the basis of a scientific world view, I don't see what is the big difference between natural selection over millions of years producing a mental operating system that is fixed at birth or natural selection proceeding over millions of years and then continuing for a few decades after birth. Neural Darwinism may rule out some reductionist theories of the mind of the sort that are based on analogies with artificial intelligence, but it does not rule out the hope of other thoroughly reductionist views of mentality.

When Edelman says that a person cannot be reduced to molecular interactions, is he saying anything different (except in degree) than a botanist or a meteorologist who says that a rose or a thunderstorm cannot be reduced to molecular interactions? It may or may not be silly to pursue reductionist programs of research on complicated systems that are strongly conditioned by history, like brains or roses or thunderstorms. What is never silly is the perspective, provided by reductionism, that apart from historical accidents these things ultimately are the way they are because of the fundamental principles of physics.

Roger Penrose strayed some time ago from his successful research in mathematical physics to think about the mind. As in his earlier books, he argued at the Jesus College symposium that "classical [that is, pre-quantum] physics seems incapable of explaining a phenomenon so deeply mysterious as consciousness." I gather that Edelman agrees with Penrose because he finds the determinism of classical physics uncongenial. Determinism is logically distinct from reductionism, but the two doctrines tend to go together because the reductionist goal of explanation is tied in with the determinist idea of prediction; we test our explanations by their power to make successful predictions. This must be what Edelman means when he speaks of modern physics (that is, quantum mechanics) as refuting Enlightenment ideas of reductionism.

Of course, everything is at bottom quantum mechanical; the question is whether quantum mechanics will appear *directly* in the theory of the mind, and not just in the deeper-level theories like chemistry on which the theory of the mind will be based. Edelman and Penrose might be right about this, but I doubt it. It is precisely those systems that can be approximately described by pre-quantum classical mechanics whose extreme sensitivity to initial conditions makes their behavior unpredictable for all practical purposes. In quantum mechanics isolated systems are governed by an equation (the Schroedinger equation) whose solutions are strictly speaking fully deterministic, never chaotic. The famous uncertainties in the positions and velocities of particles discovered by Heisenberg do not arise in isolated systems; they arise only when we let such a system interact with a measuring apparatus.

There is another reason for some of the opposition to reductionism, and specifically to the perspective provided by grand reductionism. It is that this perspective removes much of the traditional motivation for belief in God. This is especially true, for example, of one of the great reductionist episodes in the history of science: first Darwin and Wallace explained biological evolution as a consequence of heredity and natural selection; then twentieth-century biologists explained heredity as a result of genes and mutations; and then Francis Crick and James Watson explained the genetic mechanism as a consequence of the structure of the DNA molecule, which with a large enough computer could be explained as a solution of the Schroedinger equation. Václav Havel has described the corrosion of religious faith as a reason for his own reservations about much of science. In a 1987 article[5] he complained that modern science "abolishes as mere fiction even the innermost foundation of our natural world; it kills God and takes his place on the vacant throne." Then, last year, in a widely quoted speech, he added that "we may know immeasurably more about the universe than our ancestors did, and yet it increasingly seems that they knew something more essential about it than we do, something that escapes us."[6]

On the other hand, some people are attracted to reductionist science for the same reason. This is an old story. Thales' ocean had no room for Poseidon. In Hellenistic times the cult leader Epicurus

5. "Politics and Conscience," in *Václav Havel, or Living in Truth* (London: Faber and Faber, 1987), p. 138. I should add that Havel approves of some aspects of modern science: the anthropic principle and the Gaia hypothesis. This is cold comfort to the working scientist; Havel misunderstands the anthropic principle and overrates the Gaia hypothesis.

6. Speech at Independence Hall, Philadelphia, July 4, 1994; excerpted in *The New York Times*, July 8, 1994, p. A27.

adopted the atomistic theory of Democritus as an antidote to belief in the Olympian gods. I think that Midgley is absolutely right
in arguing that scientists are often driven in their work by motives
of this sort. Of course, none of this bears on the question of
whether the reductionist perspective is correct. And since in fact it
is correct, we had all better learn to live with it.

There is one limitation on the scientific world view that I am
glad to acknowledge. Science may be able to tell us how to explain
or to get what we value, but it can never tell us what we ought to
value. Moral or aesthetic statements are simply not of the sort
which it is appropriate to call true or false. I think Midgley would
agree, but I am not sure whether Atkins would, and certainly
many others would not. According to the British press, the Bishop
of Edinburgh recently argued that, since people are genetically
preconditioned toward adultery, the Church should not condemn
it. Whatever you think of adultery, it is simply a non sequitur to
draw moral lessons from genetics. Ronald Reagan made the same
silly mistake when he argued that abortion should he banned because science has not yet decided whether the fetus is alive. Whatever definition of life scientists may find convenient, and at whatever point in pregnancy a fetus may start to match that definition,
the question of the value we should place on (say) a newly fertil-
ized human egg is one that is entirely open to individual moral
judgment. (Not that this is the only issue in the debate over abortion, or even the one that necessarily motivates opponents of abortion.) Science can't even justify science; the decision to explore the
world as it is shown to us by reason and experiment is a moral
one, not a scientific one.

None of the participants in the symposium at Jesus College
seems to have addressed the really urgent problem confronting
reductionism: *Is it worth what it costs?* After all, there are many
competing reasons for doing science. Some research (e.g., medicine, much of chemistry) is done for practical purposes, or for use
in other fields. Some of it (e.g., medicine again, especially psychiatry and psychology, human evolution) is done because we are nat-

urally interested in ourselves. Some of it deals with things (e.g., the mind again, black holes, superconductivity) that are so weird and impressive that we can't help trying to understand them. Some research is done because we suspect the phenomena that we study (e.g., superconductivity again, turbulence, sex ratios in animal populations) will have explanations that are mathematically beautiful. All of these types of research compete for funds with research that is done because it moves us closer to the reductionist goal of finding the laws of nature that lie at the starting point of all chains of explanation.

The problem facing science is not (as most of the Jesus College symposiasts seem to think) that the reductionist imperative is putting the rest of science at risk. Few if any of us who are interested in the search for the laws of nature doubt the validity of the other motives for research. (I suspect that eventually I will come to feel that research on cancer or heart disease is more important than anything else.) The problem is that some people, including some scientists, deny that the search for the final laws of nature has its own special sort of value, a value that *also* should be taken into account in deciding how to fund research.

At present, the search for final explanations takes place chiefly in elementary particle physics. But research on elementary particles has become very expensive, because the laws of nature are revealed more clearly in the collisions of particles in high energy accelerators than in what is going on around us in everyday life. Cosmology is also important here. As John Barrow reminded the symposium, in order to understand the world we need to know not only the laws of nature but also the initial conditions. Some theorists hope that the initial conditions may ultimately he derived from the laws of nature, but we are a long way from that goal. Cosmological research too is very expensive, requiring observatories like Hubble and COBE and AXAF that are carried above the earth's atmosphere on artificial satellites.

For budgetary reasons this sort of research is slowly coming to a halt in the United States. The Superconducting Super Collider project was canceled partly because of arguments that such research is best done at existing laboratories or through international collaboration; but the same Congress that killed the Super Collider went on to cut research funding at other national laboratories, and the present Congress has shown no eagerness to cooperate with Europe in building its next large accelerator near Geneva, the Large Hadron Collider.

In the debates over these funding decisions, an important part was and is played by scientists, including some physicists, who oppose spending for elementary particle physics. In part, these scientists take this position because they hope to see this money spent on research in their own fields, a hope that was disappointed when the funds saved by canceling the Super Collider disappeared into the general budget. But also at work is a perfectly sincere lack of appreciation of the reductionist tradition in science, a tradition that in our time is embodied in the physics of elementary particles and fields. It is good that reductionism is discussed by talented people like those who met at Jesus College, but I wish that their discussion could have been more to the point.

11

Physics and History

Gerald Holton is Mallinckrodt Professor of Physics and Professor of History of Science emeritus at Harvard. I first met him when I came to Cambridge from Berkeley in 1966, and we became good friends. More times than I can tell, when I had to find out something about the history of science, I turned to Gerry for help, and was never disappointed. So naturally I accepted when I was asked to speak at an October 1996 meeting in Gerry's honor at the American Academy of Arts and Sciences. The talks given at this meeting were published in *Daedalus,* the journal of the American Academy that Gerry had launched as a quarterly four decades earlier.

In this talk there is a description of some debates that I had had with Harry Collins and other historians of science, about the use of scientific knowledge in writing the history of science. This debate has been continued in a book edited by Collins and Jay Labinger, to be published by the University of Chicago Press.

I am one of the few contributors to this issue of *Daedalus* who is not in any sense a historian. I work and live in the country of physics, but history is the place that I love to visit as a tourist. Here I wish to consider, from the perspective of a physicist, the uses that history has for physics and that physics has for history, and the dangers that each poses to the other.

I should begin by observing that one of the best uses of the history of physics is to help physicists teach physics to nonphysicists. You know, although many nonphysicists are nice people, they are rather odd. Physicists get a natural pleasure out of being able to calculate all sorts of things, everything from the shape of a cable in a suspension bridge to the flight of a projectile or the energy of the hydrogen atom. Nonphysicists, for some reason, do not appear to experience a comparable thrill in considering such matters. This is sad but true. It poses a problem, because if one intends to teach nonphysicists the machinery by which these calculations are done, one is simply not going to get a very receptive class.

History offers a way around this pedagogical problem. Everyone loves a story. For example, a professor can tell the story (as I did in a book and in courses at Harvard and the University of Texas) of the discovery of the subatomic particles—the electron, the proton, and all the others.[1] In the course of learning this history, students—in order to understand what was going on in the laboratories of J. J. Thomson, Ernest Rutherford, and our other heroes—have to learn something about how particles move under the influence of various forces, about energy and momentum, and about electric and magnetic fields. Thus, in order to understand the stories, they need to learn some of the physics we think they should know. It was Gerald Holton's 1952 book *Introduction to Concepts and Theories of Physical Science* that first utilized precisely this method of teaching physics; Holton told the story of the development of modern physics, all the while using it as a vehicle for teaching physics. Unfortunately, despite his efforts and those of many who came after him, the problem of teaching physics to nonphysicists remains unsolved. It is still one of the great problems facing education—how to communicate "hard sciences" to an unwilling public. In many colleges throughout the country the effort has been given up completely. Visiting small liberal arts colleges, one often finds that the only course offered in physics at all is the usual course for premedical students. Many undergraduates will thus never get the chance to encounter a hook like Holton's.

1. Steven Weinberg, *The Discovery of Subatomic Particles* (San Francisco: Scientific American/Freeman, 1982).

History plays a special role for elementary particle physicists like myself. In a sense, our perception of history resembles that of Western religions, Christianity and Judaism, as compared to the historical view of other branches of science, which are more like that of Eastern religious traditions. Christianity and Judaism teach that history is moving toward a climax—the day of judgment; similarly, many elementary particle physicists think that our work in finding deeper explanations of the nature of the universe will come to an end in a final theory toward which we are working. An opposing perception of history is held by those faiths that believe that history will go on forever, that we are bound to the wheel of endless reincarnation. Particle physicists' vision of history is quite different from that of most other scientists. They look forward to an endless future of finding interesting problems—understanding consciousness, or turbulence, or high-temperature superconductivity—that will go on forever. In elementary particle physics our aim is to put ourselves out of business. This gives a historical dimension to our choice of the sort of work on which to concentrate. We tend to seek out problems that will further this historical goal—not just work that is interesting, useful, or that influences other fields, but work that is historically progressive, that moves us toward the goal of a final theory.

In this quest for a final theory, problems get bypassed. Things that once were at the frontier, as nuclear physics was in the 1930s, no longer are. This has happened recently to the theory of strong interactions. We now understand the strong forces that hold the quarks together inside the nuclear particles in terms of a quantum field theory called quantum chromodynamics. When I say that we understand these forces, I do not mean that we can do every calculation we might wish to do; we are still unable to solve some of the classic problems of strong interaction physics, such as calculating the mass of the proton (the nucleus of the hydrogen atom). A silly letter in *Physics Today* recently asked why we bother to talk about speculative fundamental theories like string theory when the long-standing problem of calculating the mass of the proton remains to be solved. Such criticism misses the point of research focused on a historical goal. We have solved enough problems using quantum chromodynamics to know that the theory is right; it is not necessary to mop up all the islands of unsolved problems in order to make progress toward a final theory. Our situation is a little like that of the United States Navy in World War II: bypassing Japanese strong points like Truk or Rabaul, the navy instead moved on to take Saipan, which was closer to its goal of the Japanese home islands. We too must learn that we can bypass some problems. This is not to say that these problems are not worth working on; in fact, some of my own recent work has been in the application of quantum chromodynamics to nuclear physics. Nuclear forces present a classic problem—one on which I was eager to work. But I am not under the illusion that this work is part of the historical progress toward a final theory. Nuclear forces present a problem that remains interesting, but not as part of the historical mission of fundamental physics.

If history has its value, it has its dangers as well. The danger in history is that in contemplating the great work of the past, the great heroic ideas—relativity, quantum mechanics, and so on—we develop such reverence for them that we become unable to reassess their place in what we envision as a final physical theory.

General Relativity provides an example of this. As developed by Einstein in 1915, General Relativity appears almost logically inevitable. There was a fundamental principle, Einstein's principle of the equivalence of gravitation and inertia, which says that there is no difference between gravity and the effects of inertia (effects like centrifugal force). The principle of equivalence can be reformulated as the principle that gravity is just an effect of the curvature of space and time—a beautiful principle from which Einstein's theory of gravitation follows almost uniquely.

But there is an "almost" here. To arrive at the equations of General Relativity, Einstein in 1915 had to make an additional assumption; he had to assume that the equations of General Relativity would be of a particular type, known as second-order partial differential equations. This is not the place to explain precisely what a second-order partial differential equation is—roughly speaking, it is an equation in which appear not only things like gravitational fields, and the rates at which these things change with time and position, but also second-order rates, the rates at which the rates change. It does not include higher-order rates, for instance, third-order rates—the rates at which the rates of change of the rates of change of the fields that are changing are themselves changing.

This was a technical mathematical assumption, very different from a grand physical principle like the principle of equivalence of gravitation and inertia. It was just a limit on the sorts of equations that were to be allowed in the theory. So why did Einstein make this assumption?

For one thing, people were used to such equations at the time: the equations of Maxwell that govern electromagnetic fields and the wave equations that govern the propagation of sound are all second-order differential equations. For a physicist in 1915, therefore, it was a natural assumption. If a theorist does not know what else to do, it is a good tactic to assume the simplest possibility; this is more likely to produce a theory that one can actually solve, providing at least the chance to decide whether or not it agrees with experiment. In Einstein's case, the tactic worked.

But this kind of pragmatic success does not in itself provide a rationale that would satisfy, of all people, Einstein. Einstein's goal was never simply to find theories that fit the data. Remember, it was Einstein who said that the purpose of the kind of physics he did was "not only to know how nature is and how her transactions are carried through, but also to reach as far as possible the Utopian and seemingly arrogant aim of knowing why nature is thus and not otherwise." He certainly was not achieving that goal when he arbitrarily assumed that the equations for General Relativity were second-order differential equations. He could have made them fourth-order differential equations, but he did not.

Our perspective on this today, which has been developing gradually over the last fifteen or twenty years, is different from that of Einstein. Many of us now regard General Relativity as nothing but an effective field theory—that is to say, a field theory that provides an approximation to a more fundamental theory, an approximation valid in the limit of large distances, probably including any distances that are larger than the scale of an atomic nucleus. Indeed, if one supposes that there really are terms in the Einstein equations that involve rates of fourth or higher order, such terms would still play no significant role at sufficiently large distances. This is why Einstein's tactic worked. There is a rational reason for assuming the equations are second-order differential equations, which is that any terms in the equations involving higher-order rates would not make much of a difference in any astronomical observations. As far as I know, however, this was not Einstein's rationale.

This may seem rather a minor point to raise here, but in fact the most interesting work today in the study of gravitation is precisely in contexts in which the presence of higher-order rates in the field equations would make a big difference. The most important problem in the quantum theory of gravity arises from the fact that when one does various calculations—as, for instance, in attempting to calculate the probability that a gravitational wave will be scattered by another gravitational wave—one gets answers that turn out to be infinite. Another problem arises in the classical theory of gravitation from the presence of singularities: matter can apparently collapse to a point in space with infinite energy density and infinite spacetime curvature. These absurdities, which have been exercising the attention of physicists for many decades, are precisely problems that involve gravity at very short distances—not the large distances of astronomy, but distances much smaller than the size of an atomic nucleus.

From the point of view of modern effective field theory, there are no infinities in the quantum theory of gravity. The infinities are canceled in exactly the same way that they are in all our other theories, by just being absorbed into a redefinition of parameters in the field equations; but this works only if we include terms in the equations involving rates of fourth order and all higher orders, not just the terms in the equations of Einstein's original theory. (John Donaghue of the University of Massachusetts at Amherst has done more than anyone in showing how this works.) The old problems of infinities and singularities in the theory of gravitation cannot be dealt with by taking Einstein's original theory seriously as a fundamental theory. From the modern point of view—if you like, from my point of view—Einstein's theory is nothing but an approximation valid at long distances, which cannot be expected to deal successfully with infinities and singularities.

Yet some professional quantum gravitationalists (if that is the word) spend their whole careers studying the applications of the original Einstein theory, the one that only involves second-order differential equations, to problems involving infinities and singularities. Elaborate formalisms have been developed that aim to look at Einstein's theory in a more sophisticated way, in the hope that doing so will somehow or other make the infinities or singularities go away. This ill-placed loyalty to General Relativity in its original form persists because of the enormous prestige the theory earned from its historic successes.

But it is precisely in this manner that the great heroic ideas of the past can weigh upon us, preventing us from seeing things in a fresh light. It is those ideas that were most successful of which we should be most wary. Otherwise we become like the French army, which in 1914 tried to imitate the successes of Napoleon and almost lost the war—and then in 1940 tried to imitate the 1916 success of Marshal Philippe Pétain in defending Verdun, only to suffer decisive defeat. Such examples exist in the history of physics as well. For instance, there is an approach to quantum field theory called second quantization, which fortunately no longer plays a significant role in research but continues to play a role in the way that textbooks are written. Second quantization goes back to a paper written in 1927 by Jordan and Oscar Klein that put forth the idea that after one has introduced a wave function in quantizing a theory of particles, one should then quantize the wave function. Surprisingly, many people still think that this is the way to look at quantum field theory, though it is not.

We have to expect the same fate for our present theories. The Standard Model of weak, electromagnetic, and strong forces, which is used to describe nature under conditions that can be explored in today's accelerators, is not likely to disappear or be proved wrong, but it has already come to be looked at in quite a different way than it was twenty years ago. Most particle physicists now think of the Standard Model as only an effective field theory that provides a low-energy approximation to a more fundamental theory.

Enough about the danger of history to science; let me now take up the danger of scientific knowledge to history. This arises from a tendency to imagine that discoveries are made according to our present understandings. Gerald Holton has done as much as anyone in trying to point out these dangers and puncture these misapprehensions. In his papers about Einstein he shows, for example, that the natural deduction of the Special Theory of Relativity from the experiment of Michelson and Morley, which demonstrated that there is no motion through the ether, is not at all the way Einstein actually came to Special Relativity.

Holton has also written about Kepler. At one point in my life I was one of those people who thought that Kepler deduced his famous three laws of planetary motion solely by studying the data of the Danish astronomer Tycho Brahe. But Holton points out how much else besides data, how much of the spirit of the Middle Ages and of the Greek world, went into Kepler's thinking—how many things that we now no longer associate with planetary motion were on Kepler's mind. By assuming that scientists of the past thought about things the way we do, we make mistakes about history; what is worse, we lose appreciation for the difficulties, for the intellectual challenges, that they faced.

Once, at the Tate Gallery in London, I overheard a lecturer talking to a tour group about the Turner paintings. J. M. W. Turner was very important, said the guide, because he foreshadowed the Impressionists of the later nineteenth century. I had thought Turner was an important painter because he painted beautiful pictures; Turner did not know that he was foreshadowing anything. One has to look at things as they really were in their own time.

This also applies, of course, to political history. There is a term "Whig interpretation of history," which was invented by Herbert Butterfield in a lecture in 1931. As Butterfield explained it, "The Whig historian seems to believe that there is an unfolding logic in history." He went on to attack the person he regarded as the archetypal Whig historian, Lord Acton, who wished to use history as a way to pass moral judgments on the past. Acton wanted history to serve as the "arbiter of controversy, the upholder of that moral standard which the powers of earth and religion itself tend constantly to depress. . . . It is the office of historical science to maintain morality as the sole impartial criterion of men and things." Butterfield went on to say:

> If history can do anything it is to remind of us of those complications that undermine our certainties, and to show us that all our judgments are merely relative to time and circumstance. . . . We can never assert that history has proved any man right in the long run. We can never say that the ultimate issue, the succeeding course of events, or the lapse of time have proved that Luther was right against the pope or that Pitt was wrong against Charles James Fox.[2]

This is the point at which the historian of science and the historian of politics should part company. The passage of time has shown that, for example, Darwin was right against Lamarck, the atomists were right against Ernst Mach, and Einstein was right against the experimentalist Walter Kaufmann (who had presented data contradicting Special Relativity). To put it another way, Butterfield was correct in his comments on political and social history;

2. Herbert Butterfield, *The Whig Interpretation of History* (New York: Scribners, 1951), p. 75.

there is no sense in which Whig morality (much less the Whig Party) existed at the time of Luther. But nevertheless it is true that natural selection was working during the time of Lamarck, and the atom did exist in the days of Mach, and fast electrons behaved according to the laws of relativity even before Einstein. Present scientific knowledge has the potentiality of being relevant in the history of science in the way that present moral and political judgments may not be relevant in political or social history.

Many historians, sociologists, and philosophers of science have taken the desire for historicism, the worry about falling into a Whig interpretation of history, to extremes. To quote Holton, "Much of the recent philosophical literature claims that science merely staggers from one fashion, conversion, revolution, or incommensurable exemplar to the next in a kind of perpetual, senseless Brownian motion, without discernible direction or goal."[3] I made a similar observation in an address to the American Academy of Arts and Sciences about a year and a half ago, noting in passing that there are people who see scientific theories as nothing but social constructions. The talk was circulated by the Academy, as is its practice, and a copy of it fell into the hands of someone who over twenty years ago had been closely associated with a development known as the Sociology of Scientific Knowledge (SSK). He wrote me a long and unhappy letter; among other things, he complained about my remark that the Strong Program initiated at the University of Edinburgh embodied a radical social constructivist view, in which scientific theories are nothing but social constructions. He sent me a weighty pile of essays, saying that they demonstrated that he and his colleagues do recognize that reality plays a role in our world. I took this criticism to heart and decided that I would read the essays. I also looked back over some old correspondence that I had had with Harry Collins, who for many years led the well-known Sociology of Scientific Knowledge group

3. Gerald Holton, *Einstein, History, and Other Passions* (Reading, Mass.: Addison-Wesley, 1996), p. 22.

at the University of Bath. My purpose in all of this was to look at these materials from as sympathetic a point of view as I could, try to understand what they were saying, and assume that they must be saying something that is not absurd.

I did find described (though not espoused) in an article by David Bloor, who is one of the Edinburgh group, and also in my correspondence with Harry Collins, a point of view that on the face of it is not absurd. As I understand it, there is a position called "methodological idealism" or "methodological antirealism," which holds that historians or sociologists should take no position on what is ultimately true or real. Instead of using today's scientific knowledge as a guiding principle for their work, the argument goes, they should try to look at nature as it must have been viewed by the scientists under study at the time that those scientists were working. In itself, this is not an absurd position. In particular, it can help to guard us against the kind of silliness that (for instance) I was guilty of when I interpreted Kepler's work in terms of what we now know about planetary motion.

Even so, the attitude of methodological antirealism bothered me, though for a while I could not point to what I found wrong with it. In preparing this essay I have tried to think this through, and I have come to the conclusion that there are a number of minor things wrong with methodological antirealism: it can hamper historical research, it is often boring, and it is basically impossible. More significant, however, it has a major drawback—in an almost literal sense, it misses the point of the history of science.

Let me first address the minor points. If it were really possible to reconstruct everything that happened during some past scientific discovery, then it might he helpful to forget everything that has happened since; but in fact much of what occurred will always be unknown to us. Consider just one example. J. J. Thomson, in the experiments that made him known as the discoverer of the electron, was measuring a certain crucial quantity, the ratio of the electron's mass to its charge. As always happens in experimental work, he found a range of values. Although he quoted various values in his published work, the values he would always refer to as his favorite results were those at the high end of the range. Why did Thomson quote the high values as his favorite values? It is possible that Thomson knew that on the days those results had been obtained he had been more careful; perhaps those were the days he had not bumped into the laboratory table, or before which he'd had a good night's sleep. But the possibility also exists that perhaps his first values had been at the high end of the range, and he was determined to show that he had been right at the beginning. Which explanation is correct? There is simply no way of reconstructing the past. Not his notebooks, not his biography—nothing will allow us now to reconstitute those days in the Cavendish Laboratory and find out on which days Thomson was more clumsy or felt more sleepy than usual. There is one thing that we do know, however: the actual value of the ratio of the electron's mass to its charge, which was the same in Thomson's time as it is in our own. We know, in fact, that the actual value is not at the high end but, rather, at the low end of the range of Thomson's experimental values, which strongly suggests that when Thomson's measurements gave high values they were not actually more careful—and that therefore it is more likely that Thomson quoted these values because he was trying to justify his first measurements.

This is a trivial example of the use of present scientific knowledge in the history of science, because here we are just talking about a number, not a natural law or an ontological principle. I chose this example simply because it shows so clearly that to decide to ignore present scientific knowledge is often to throw away a valuable historical tool.

A second minor drawback of methodological antirealism is that a reader who does not know anything about our present understanding of nature is likely to find the history of science terribly boring. For instance, a historian might describe how in 1911 the Dutch physicist Kamerlingh Onnes was measuring the electrical resistance of a sample of cold mercury and thought that he had found a short circuit. The historian could go on for pages and pages describing how Onnes searched for the short circuit, and how he took apart the wiring and put it back together again without any success in finding the source of the short circuit. Could anything be more boring than to read this description, if one did not know in advance that there *was* no short circuit—that what Onnes was observing was in fact the vanishing of the resistance of mercury when cooled to a certain temperature, and that this was nothing less than the discovery of superconductivity? Of course, it is impossible today for a physicist or a historian of physics not to know about superconductivity. Indeed, we are quite incapable while reading about the experiments of Kamerlingh Onnes of imagining that his problem was nothing but a short circuit. Even if one had never heard of superconductivity, the reader would know that there was something going on besides a short circuit; why else would the historian bother with these experiments? Plenty of experimental physicists have found short circuits, and no one studies them.

But these are minor issues. The main drawback of methodological antirealism is that it misses the point about the history of science that makes it different from other kinds of history: Even though a scientific theory is in a sense a social consensus, it is unlike any other sort of consensus in that it is culture-free and permanent.

This is just what many sociologists of science deny. David Bloor stated in a talk at Berkeley a year ago that "the important thing is that reality underdetermines the scientists' understanding." I gather he means that although he recognizes that reality has some effect on what scientists do—so that scientific theories are not "nothing but" social constructions—scientific theories are also not what they are simply because that is the way nature is. In a similar spirit, Stanley Fish, in a recent article in *The New York Times,* argued that the laws of physics are like the rules of baseball. Both are certainly conditioned by external reality—after all, if baseballs moved differently under the influence of earth's gravity, the rules would call for the bases to be closer together or farther apart—but the rules of baseball also reflect the way that the game developed historically and the preferences of players and fans.[4]

Now, what Bloor and Fish say about the laws of nature may apply while these laws are being discovered. Holton's work on Einstein, Kepler, and superconductivity has shown that many cultural and psychological influences enter into scientific work. But the laws of nature are not like the rules of baseball. They are culture-free and they are permanent—not as they are being developed, not as they were in the mind of the scientist who first discovers them, not in the course of what Bruno Latour and Steve Woolgar call "negotiations" over what theory is going to be accepted, but in their final form, in which cultural influences are refined away. I will even use the dangerous words "nothing but": aside from inessentials like the mathematical notation we use, the laws of physics as we understand them now are nothing but a description of reality.

I cannot prove that the laws of physics in their mature form are culture-free. Physicists live embedded in the Western culture of the late twentieth century, and it is natural to be skeptical if we say that our understanding of Maxwell's equations, quantum mechanics, relativity, or the Standard Model of elementary particles is culture-free. I am convinced of this because the purely scientific arguments for these theories seem to me overwhelmingly convincing. I can add that as the typical background of physicists has changed, in particular as the number of women and Asians in physics has increased, the nature of our understanding of physics has not changed. These laws in their mature form have a toughness that resists cultural influence.

The history of science is further distinguished from political or artistic history (in such a way as to reinforce my remarks about the influence of culture) in that the achievements of science become permanent. This assertion may seem to contradict a statement at

4. Stanley Fish, "Professor Sokal's Bad Joke," *The New York Times*, May 21, 1996, op-ed section.

the beginning of this essay—that we now look at General Relativity in a different way than Einstein did, and that even now we are beginning to look at the Standard Model differently than we did when it was first being developed. But what changes is our understanding of both why the theories are true and their scope of validity. For instance, at one time we thought there was an exact symmetry in nature between left and right, but then it was discovered that this is only true in certain contexts and to a certain degree of approximation. But the symmetry between right and left was not a simple mistake, nor has it been abandoned; we simply understand it better. Within its scope of validity, this symmetry has become a permanent part of science, and I cannot see that this will ever change.

In holding that the social constructivists missed the point, I have in mind a phenomenon known in mathematical physics as the approach to a fixed point. Various problems in physics deal with motion in some sort of space. Such problems are often governed by equations dictating that wherever one starts in the space, one always winds up at the same point, known as a fixed point. Ancient geographers had something similar in mind when they said that all roads lead to Rome. Physical theories are like fixed points, toward which we are attracted. Starting points may be culturally determined, paths may be affected by personal philosophies, but the fixed point is there nonetheless. It is something toward which any physical theory moves; when we get there we know it, and then we stop.

The kind of physics I have done for most of my life, working in the theory of fields and elementary particles, is moving toward a fixed point. But this fixed point is unlike any other in science. The final theory toward which we are moving will be a theory of unrestricted validity, a theory applicable to all phenomena throughout the universe—a theory that, when finally reached, will be a permanent part of our knowledge of the world. Then our work as elementary particle physicists will be done, and will become nothing but history.

12

Sokal's Hoax

———

In May 1996 I read a story in *The New York Times* about a delightful academic hoax played by the physicist Alan Sokal. He had submitted an article to *Social Text*, a journal of trendy cultural studies, and then, after it was published, he revealed that he had intentionally stuffed his article with pretentious nonsense. I knew Sokal slightly; as an undergraduate at Harvard he had taken the graduate-level course on quantum field theory that I had given in 1974–75, and he had gone on to do first-rate work in mathematical physics, some of which I had quoted in my treatise on quantum field theory. I was curious to see what sort of mischief Sokal had been up to, so I read his article in *Social Text*. It seemed to me that Sokal had done a great service in exposing and satirizing the failings of those postmoderns and cultural relativists whom he had quoted. I wrote to Robert Silvers, editor of *The New York Review of Books*, proposing an article on Sokal's hoax; the essay below was the result.

A good number of Sokal's targets were French intellectuals. As quoted and parodied by Sokal, they were shown not only to be using arguments and examples from modern physics and mathematics that they clearly did not understand, but also to delight in wordy obscurity. In my essay I sided with Sokal on both points, and in particular came down hard on Jacques Derrida. It is not only physicists who take this negative view of the writing of some leading French intellectuals. The charge of obscurity had been made before Sokal, with special force in a 1989 article[1] by John

---

1. John Weightman, "On Not Understanding Michel Foucault," *The American Scholar* 58 (Summer 1989), 383.

Weightman, emeritus professor of French at the University of London. Weightman, a veteran Francophile, bewailed the replacement of the old tradition of clarity in French writing—"Ce qui n'est pas clair n'est pas français"—with one that was deliberately arcane—"Ce qui n'est pas un peu obscur n'est plus vraiment parisien." And he blamed this pettifoggery on the example set by Roland Barthes, Jacques Lacan, Michel Foucault, and Jacques Derrida.

As it turned out, my essay went beyond the issues addressed by Sokal, going so far as to cast doubt on the existence of any cultural implications of twentieth-century discoveries in physical science. This has attracted a good deal of angry comment, some of which is discussed in the next essay in this collection.

Like many other scientists, I was amused when I heard about the prank played by the NYU mathematical physicist Alan Sokal, who late in 1994 submitted a sham article to the cultural studies journal *Social Text*. In the article Sokal reviewed several current topics in physics and mathematics, and, tongue in cheek, drew various cultural, philosophical, and political morals that he felt would appeal to fashionable academic commentators who question the claims of science to objectivity.

The editors of *Social Text* did not detect that Sokal's article was a hoax, and they published it in the journal's Spring/Summer 1996 issue.[2] The hoax was revealed by Sokal himself in an article for another journal, *Lingua Franca*,[3] in which he explained that his *Social Text* article had been "liberally salted with nonsense," and in his opinion was accepted only because "(a) it sounded good and (b) it flattered the editors' ideological preconceptions." Newspapers and newsmagazines throughout the United States and Brit-

2. Alan D. Sokal, "Transgressing the Boundaries: Toward a Transformative Hermeneutics of Quantum Gravity," *Social Text* (Spring/Summer 1996), pp. 217–252.

3. Alan D. Sokal, "A Physicist Experiments with Cultural Studies," *Lingua Franca* (May/June 1996), pp. 62–64.

ain carried the story, and Sokal's hoax appeared likely to join the small company of legendary academic hoaxes, along with the pseudo-fossils of Piltdown man planted by Charles Dawson and the pseudo-Celtic epic *Ossian* written by James Macpherson. The difference is that Sokal's hoax served a public purpose, to attract attention to what Sokal saw as a decline of standards of rigor in the academic community, and for that reason it was disclosed immediately by the author himself.

The targets of Sokal's satire occupy a broad intellectual range. There are those "postmoderns" in the humanities who like to surf through avant-garde fields like quantum mechanics or chaos theory to dress up their own arguments about the fragmentary and random nature of experience. There are those sociologists, historians, and philosophers who see the laws of nature as social constructions. There are cultural critics who find the taint of sexism, racism, colonialism, militarism, or capitalism not only in the practice of scientific research but even in its conclusions. Sokal did not satirize creationists or other religious enthusiasts who in many parts of the world are the most dangerous adversaries of science,[4] but his targets were spread widely enough, and he was attacked or praised from all sides.

Entertaining as this episode was, I could not immediately judge from press reports what it proved. Suppose that, with tongue in cheek, an economist working for a labor union submitted an article to *The National Review,* giving what the author thought were false economic arguments against an increase in the statutory minimum wage. What would it prove if the article was accepted for publication? The economic arguments might still be cogent, even though the author did not believe in them.

I thought at first that Sokal's article in *Social Text* was intended to be an imitation of academic babble, which any editor should

4. In an afterword, "Transgressing the Boundaries," submitted to *Social Text,* Sokal explained that his goal was not so much to defend science as to defend the left from postmodernists, social constructivists, and other trendy leftists.

have recognized as such. But in reading the article I found that this is not the case. The article expresses views that I find absurd, but with a few exceptions Sokal at least makes it pretty clear what these views are. The article's title, "Transgressing the Boundaries: Toward a Transformative Hermeneutics of Quantum Gravity," is more obscure than almost anything in his text. (A physicist friend of mine once said that in facing death he drew some consolation from the reflection that he would never again have to look up the word "hermeneutics" in the dictionary.) In fact I got the impression that Sokal finds it difficult to write unclearly.

Where the article does degenerate into babble, it is not in what Sokal himself has written, but in the writings of the genuine postmodern cultural critics he quotes. Here, for instance, is a quote that he takes from the oracle of deconstruction, Jacques Derrida:

> The Einsteinian constant is not a constant, is not a center. It is the very concept of variability: it is, finally, the concept of the game. In other words, it is not the concept of *something*—of a center starting from which an observer could master the field— but the very concept of the game.

I have no idea what this is intended to mean.

I suppose that it might be argued that articles in physics journals are also incomprehensible to the uninitiated. But we physicists are forced to use a technical language, the language of mathematics. Within this limitation, we try to be clear, and when we fail we do not expect our readers to confuse obscurity with profundity. It never was true that only a dozen people could understand Einstein's papers on General Relativity, but if it had been true, it would have been a failure of Einstein's, not a mark of his brilliance. The papers of Edward Witten of the Institute for Advanced Study at Princeton, which are today consistently among the most significant in the promising field of string theory, are notably easier for a physicist to read than most other work in string theory. In contrast, Derrida and other postmoderns do not seem to be saying anything that requires a special technical language, and they do not seem to be trying very hard to be clear. But those who admire such writings presumably would not have been embarrassed by Sokal's quotations from them.

Part of Sokal's hoax was his description of developments in physics. Much of his account was quite accurate, but it was heavily adulterated with howlers, most of which would have been detected by any undergraduate physics major. One of his running jokes had to do with the word "linear." This word has a precise mathematical meaning, arising from the fact that certain mathematical relationships are represented graphically by a straight line.[5] But for some postmodern intellectuals, "linear" has come to mean unimaginative and old-fashioned, while "nonlinear" is understood to be somehow perceptive and avant-garde. In arguing for the cultural importance of the quantum theory of gravitation, Sokal refers to the gravitational field in this theory as "a noncommuting (and hence nonlinear) operator." Here "hence" is ridiculous, "noncommuting"[6] does not imply "nonlinear," and in fact quantum mechanics deals with things that are both noncommuting and linear.

Sokal also writes that "Einstein's equations [in the General Theory of Relativity] are highly nonlinear, which is why traditionally trained mathematicians find them so difficult to solve." The joke is in the words "traditionally trained." Einstein's equations are nonlinear, and this does make them hard to solve; but they are hard for anyone to solve, especially someone who is not traditionally

5. For instance, there is a linear relation between the number of calories in a cake and the amounts of each of the various ingredients: the graph of calories versus ounces of any one ingredient, when we hold the amounts of all the other ingredients fixed, is a straight line. In contrast, the relation between the diameter of a cake (of fixed height) and the amounts of its ingredients is not linear.

6. Operations are said to be noncommuting if the result when you perform several of them depends on the order in which they are performed. For instance, rotating your body by, say, thirty degrees around the vertical axis and then rotating it by thirty degrees around the north-south direction leaves you in a different position than these operations would if they were carried out in the opposite order. Try it and see.

trained. Continuing with General Relativity, Sokal correctly remarks that its description of curved spacetime allows arbitrary changes in the spacetime coordinates that we use to describe nature. But he then solemnly pronounces that "the pi of Euclid and the G of Newton, formerly thought to be constant and universal, are now perceived in their ineluctable historicity." This is absurd; the meaning of a mathematically defined quantity like pi cannot be affected by discoveries in physics, and in any case both pi and *G* continue to appear as universal constants in the equations of General Relativity.

In a different vein, Sokal pretends to give serious consideration to a crackpot fantasy known as the "morphogenetic field." He refers to complex number theory as a "new and still quite speculative branch of mathematical physics," while in fact it is nineteenth-century mathematics and has been as well established as anything ever is. He even complains (echoing the sociologist Stanley Aronowitz) that all of the graduate students in solid state physics will be able to get jobs in that field, which will be news to many of them.

Sokal's revelation of his intentional howlers drew the angry response that he had abused the trust of the editors of *Social Text* that they had placed in his credentials as a physicist, a complaint made by both the sociologist Steve Fuller and the English professor Stanley Fish.[7] (Fish is the executive director of Duke University Press, which publishes *Social Text,* and is reputed to be the model for Morris Zapp, the master of the academic game in David Lodge's comic novels.) The editors of *Social Text* also offered the excuse that it is not a journal in which articles are submitted to experts for evaluation, but a journal of opinion.[8] Maybe under these circumstances Sokal was naughty in letting the editors rely on his

7. Steve Fuller, letter to *The New York Times,* May 23, 1996, p. 28, and Stanley Fish, "Professor Sokal's Bad Joke," op-ed article in *The New York Times,* May 21, 1996, p. 23.

8. Bruce Robbins and Andrew Ross, "Mystery Science Theater," *Lingua Franca* (July/August 1996).

sincerity, but the article would not have been very different if Sokal's account of physics and mathematics had been entirely accurate. What is more revealing is the variety of physics and mathematics bloopers in remarks by others that Sokal slyly quotes with mock approval. Here is the philosopher Bruno Latour on Special Relativity:

> How can one decide whether an observation made in a train about the behavior of a falling stone can be made to coincide with the observation of the same falling stone from the embankment? If there are only one, or even *two,* frames of reference, no solution can he found. . . . Einstein's solution is to consider *three* actors.

This is wrong; in relativity theory there is no difficulty in comparing the results of two, three, or any number of observers. In other quotations cited by Sokal, Stanley Aronowitz misuses the term "unified field theory." The feminist theorist Luce Irigaray deplores mathematicians' neglect of spaces with boundaries, though there is a huge literature on the subject. The English professor Robert Markley calls quantum theory nonlinear, though it is the only known example of a precisely linear theory. And both the philosopher Michael Serres (a member of the Académie Française) and the arch-postmodernist Jean-François Lyotard grossly misrepresent the view of time in modern physics. Such errors suggest a problem not just in the editing practices of *Social Text* but in the standards of a larger intellectual community.

It seems to me though that Sokal's hoax is most effective in the way that it draws cultural or philosophical or political conclusions from developments in physics and mathematics. Again and again Sokal jumps from correct science to absurd implications, without the benefit of any intermediate reasoning. With a straight face, he leaps from Bohr's observation that in quantum mechanics "a complete elucidation of one and the same object may require diverse points of view which defy a unique description" to the conclusion that "postmodern science" refutes "the authoritarianism and elitism inherent in traditional science." He blithely points to catastrophe theory and chaos theory as the sort of mathematics that can lead to social and economic liberation. Sokal shows that people really do talk in this way by quoting the work of others in the same vein, including applications of mathematical topology to psychiatry by Jacques Lacan and to film criticism by Jacques-Alain Miller.

I find it disturbing that the editors of Social Text thought it plausible that a sane working physicist would take the positions satirized in Sokal's article. In their defense of the decision to publish it, the editors explain that they had judged that it was "the earnest attempt of a professional scientist to seek some kind of affirmation from postmodern philosophy for developments in his field."[9] In an introduction to the issue of Social Text in which Sokal's article appears, one of the editors mentions that "many famous scientists, especially physicists, have been mystics."[10] There may be some working physicists who are mystics, though I have never met any, but I can't imagine any serious physicist who holds views as bizarre as those that Sokal satirized. The gulf of misunderstanding between scientists and other intellectuals seems to be at least as wide as when C. P. Snow worried about it three decades ago.

After Sokal exposed his hoax, one of the editors of Social Text even speculated that "Sokal's parody was nothing of the sort, and that his admission represented a change of heart, or a folding of his intellectual resolve."[11] I am reminded of the case of the American spiritualist Margaret Fox. When she confessed in 1888 that her career of séances and spirit rappings had all been a hoax, other spiritualists claimed that it was her confession that was dishonest.

Those who seek extrascientific messages in what they think they

9. Robbins and Ross, "Mystery Science Theater."
10. Andrew Ross, "Introduction," Social Text (Spring/Summer 1996), pp. 1–13.
11. Quoted by Robbins and Ross in "Mystery Science Theater."

understand about modern physics are digging dry wells. In my view, with two large exceptions, the results of research in physics (as opposed, say, to psychology) have no legitimate implications whatever for culture or politics or philosophy. (I am not talking here about the technological applications of physics, which of course do have a huge effect on our culture, or about its use as metaphor, but about the direct logical implications of purely scientific discoveries themselves.) The discoveries of physics may become relevant to philosophy and culture when we learn the origin of the universe or the final laws of nature, but not for the present.

The first of my two exceptions to this statement is jurisdictional: discoveries in science sometimes reveal that topics like matter, space, and time, which had been thought to be proper subjects for philosophical argument, actually belong in the province of ordinary science. The other, more important exception to my statement is the profound cultural effect of the discovery, going back to the work of Newton, that nature is strictly governed by impersonal mathematical laws. Of course, it still remains for us to get the laws right, and to understand their range of validity; but as far as culture or philosophy is concerned the difference between Newton's and Einstein's theories of gravitation or between classical and quantum mechanics is immaterial.

There is a good deal of confusion about this, because quantum mechanics can seem rather eerie if described in ordinary language. Electrons in atoms do not have definite positions or velocities until these properties are measured, and the measurement of an electron's velocity wipes out all knowledge of its position. This eeriness has led Andrew Ross, one of the editors of *Social Text*, to remark elsewhere that "quantitative rationality—the normative description of scientific materialism—can no longer account for the behavior of matter at the level of quantum reality."[12] This is simply wrong. By rational processes today we obtain a complete quantitative description of atoms using what is called the "wave

12. Andrew Ross, *Strange Weather* (Verso, 1991), p. 42.

function" of the atom.[13] Once one has calculated the wave function, it can be used to answer any question about the energy of the atom or its interaction with light. We have replaced the precise Newtonian language of particle trajectories with the precise quantum language of wave functions, but as far as quantitative rationality is concerned, there is no difference between quantum mechanics and Newtonian mechanics.

I have to admit at this point that physicists share responsibility for the widespread confusion about such matters. Sokal quotes some dreadful examples of Werner Heisenberg's philosophical wanderings, as for instance: "Science no longer confronts nature as an objective observer, but sees itself as an actor in this interplay between man [sic] and nature." (Heisenberg was one of the great physicists of the twentieth century, but he could not always be counted on to think carefully, as shown by his technical mistakes in the German nuclear weapons program.)[14] More recently scientists like Ilya Prigogine[15] have claimed a deep philosophical sig-

13. In general, the wave function of any system is a list of numbers, one number for every possible configuration of the system. For a single electron in an atom, the list includes a different number for each possible position of the electron. The values of these numbers give a complete description of the state of the system at any moment. One complication is that the possible configurations of any system can be described in different ways; for instance, an electron could he described in terms of its possible velocities, rather than its possible positions (but not by both at the same time). There are well-understood rules for calculating the numbers making up the wave function in one description if we know what these numbers are in any other description. Another complication is that these numbers are complex, in the sense that they generally involve the quantity known as i, equal to the square root of minus one, as well as ordinary real numbers.

14. See Jeremy Bernstein, *Hitler's Uranium Club* (American Institute of Physics, 1995).

15. For quotes and comments, see Jean Bricmont, "Science of Chaos or Chaos in Science?" *Physicalia Magazine* 17 (1995), pp. 159–208, reprinted in *The Flight from Science and Reason* (New York Academy of Sciences, 1996). A rejoinder and response are given by Ilya Prigogine and I. Antoniou, "Science of Chaos or Chaos in Science: A Rearguard Battle," *Physicalia Magazine* 17, pp. 213–218, and Jean Bricmont, "The Last Word from the Rearguard," *Physicalia Magazine* 17, pp. 219–221.

nificance for work on nonlinear dynamics,[16] a subject that is interesting enough without the hype.

So much for the cultural implications of discoveries in science. What of the implications for science of its cultural and social context? Here scientists like Sokal find themselves in opposition to many sociologists, historians, and philosophers as well as postmodern literary theorists. In this debate, the two sides often seem to be talking past each other. For instance, the sociologists and historians sometimes write as if scientists had not learned anything about the scientific method since the days of Francis Bacon, while of course we know very well how complicated the relation is between theory and experiment, and how much the work of science depends on an appropriate social and economic setting. On the other hand, scientists sometimes accuse others of taking a completely relativist view, of not believing in objective reality. With dead seriousness, Sokal's hoax cites "revisionist studies in the history and philosophy of science" as casting doubt on the post-Enlightenment dogma that "there exists an external world, whose properties are independent of any individual human being and indeed of humanity as a whole." The trouble with the satire of this particular passage is that most of Sokal's targets deny that they have any doubt about the existence of an external world. Their belief in objective reality was reaffirmed in response to Sokal's hoax both in a letter to *The New York Times* by the editors of *Social Text*[17] and in the op-ed article by Stanley Fish.

16. Nonlinear dynamics deals with cases in which the rates of change of various quantities depend nonlinearly on these quantities. For instance, the rates of change of the pressures, temperatures, and velocities at various points in a fluid like the atmosphere depend nonlinearly on these pressures, temperatures, and velocities. It has been known for almost a century that the long-term behavior of such systems often exhibits chaos, an exquisite sensitivity to the initial condition of the system. (The classic example is the way that the flapping of a butterfly's wings can change the weather weeks later throughout the world.) For physicists, the current interest in nonlinear dynamical systems stems from the discovery of general features of chaotic behavior that can be precisely predicted.

17. Bruce Robbins and Andrew Ross, letter to *The New York Times*, May 23, 1996, p. 28.

I don't mean to say that this part of Sokal's satire was unjustified. His targets often take positions that seem to me (and, I gather, to Sokal) to make no sense if there is an objective reality. To put it simply, if scientists are talking about something real, then what they say is either true or false. If it is true, then how can it depend on the social environment of the scientist? If it is false, how can it help to liberate us? The choice of scientific question and the method of approach may depend on all sorts of extrascientific influences, but the correct answer when we find it is what it is because that is the way the world is. Nevertheless, it does no good to satirize views that your opponent denies holding.

I have run into the same sort of stumbling block myself. In an early draft of my book *Dreams of a Final Theory*[18] I criticized the feminist philosopher of science Sandra Harding (a contributor to *Social Text),* for taking a relativist position that denied the objective character of physical laws. In evidence, I quoted her as calling modern science (and especially physics) "not only sexist but also racist, classist, and culturally coercive," and arguing that "physics and chemistry, mathematics and logic, bear the fingerprints of their distinctive cultural creators no less than do anthropology and history."[19] It seemed to me that this statement could make sense only to a relativist. What is the good of claiming that the conclusions of scientific research should he friendlier to multicultural or feminist concerns if these conclusions are to be an accurate account of objective reality? I sent a draft of this section to Harding, who pointed out to me various places in her writing where she had explicitly denied taking a relativist position. I took the easy way out; I dropped the accusation of relativism, and left it to the reader to judge the implications of her remarks.

Perhaps it would clarify what is at issue if we were to talk not about whether nature is real, but about the more controversial question, whether scientific knowledge in general and the laws of physics in particular are real.

18. Pantheon, 1993.
19. Sandra Harding, *The Science Question in Feminism* (Cornell University Press, 1986), pp. 9, 250.

When I was an undergraduate at Cornell I heard a lecture by a professor of philosophy (probably Max Black) who explained that whenever anyone asked him whether something was real, he always gave the same answer. The answer was "Yes." The tooth fairy is real, the laws of physics are real, the rules of baseball are real, and the rocks in the fields are real. But they are real in different ways. What I mean when I say that the laws of physics are real is that they are real in pretty much the same sense (whatever that is) as the rocks in the fields, and not in the same sense (as implied by Fish)[20] as the rules of baseball. We did not create the laws of physics or the rocks in the field, and we sometimes unhappily find that we have been wrong about them, as when we stub our toe on an unnoticed rock, or when we find we have made a mistake (as most physicists have) about some physical law. But the languages in which we describe rocks or in which we state physical laws are certainly created socially, so I am making an implicit assumption (which in everyday life we all make about rocks) that our statements about the laws of physics are in a one-to-one correspondence with aspects of objective reality. To put it another way, I think that if we ever discover intelligent creatures on some distant planet and translate their scientific works, we will find that we and they have discovered the same laws.

There is another complication here, which is that none of the laws of physics known today (with the possible exception of the general principles of quantum mechanics) are exactly and universally valid. Nevertheless, many of them have settled down to a final form, valid in certain known circumstances. The equations of electricity and magnetism that are today known as Maxwell's equations are not the equations originally written down by Maxwell; they are equations that physicists settled on after decades of subsequent work by other physicists, notably the English scientist Oliver Heaviside. They are understood today to be an approximation that is valid in a limited context (that of weak, slowly varying

20. Fish, "Professor Sokal's Bad Joke."

electric and magnetic fields), but in this form and in this limited context they have survived for a century and may be expected to survive indefinitely. This is the sort of law of physics that I think corresponds to something as real as anything else we know. On this point, scientists like Sokal and myself are apparently in clear disagreement with some of those whom Sokal satirizes. The objective nature of scientific knowledge has been denied by Andrew Ross[21] and Bruno Latour[22] and (as I understand them) by the influential philosophers Richard Rorty and the late Thomas Kuhn,[23] but it is taken for granted by most natural scientists.

I have come to think that the laws of physics are real because my experience with the laws of physics does not seem to me to be very different in any fundamental way from my experience with rocks. For those who have not lived with the laws of physics, I can offer the obvious argument that the laws of physics as we know them work, and there is no other known way of looking at nature that works in anything like the same sense. Sarah Franklin (in an article in the same issue of *Social Text* as Sokal's hoax) challenges an argument of Richard Dawkins that in relying on the working of airplanes we show our acceptance of the working of the laws of nature, remarking that some airlines show prayer films during takeoff to invoke the aid of Allah to remain safely airborne.[24] Does Franklin think that Dawkins's argument does not apply to her? If so, would she be willing to give up the use of the laws of physics in designing aircraft, and rely on prayers instead?

21. Andrew Ross was quoted by *The New York Times* on May 18, 1996, to the effect that "scientific knowledge is affected by social and cultural conditions and is not a version of some universal truth that is the same in all times and places."

22. Bruno Latour, *Science in Action* (Harvard University Press, 1987).

23. For instance, see Thomas Kuhn, "The Road since Structure," in *PSA 1990* (Philosophy of Science Association, 1991), and "The Trouble with the Historical Philosophy of Science" (1991 lecture published by the Department of the History of Science, Harvard University, 1992).

24. Sarah Franklin, "Making Transparencies: Seeing through the Science Wars," *Social Text* (Spring/Summer 1996), pp. 141–155.

There is also the related argument that although we have not yet had a chance to compare notes with the creatures on a distant planet, we can see that on earth the laws of physics are understood in the same way by scientists of every nation, race, and—yes—gender. Some of the commentators on science quoted by Sokal hope that the participation of women or victims of imperialism will change the character of science; but as far as I can see, women and Third World physicists work in just the same way as Western white male physicists do. It might be argued that this is just a sign of the power of entrenched scientific authority or the pervasive influence of Western society, but these explanations seem unconvincing to me. Although natural science is intellectually hegemonic, in the sense that we have a clear idea of what it means for a theory to be true or false, its operations are not socially hegemonic—authority counts for very little.

From time to time distinguished physicists who are past their best years, like Heisenberg in Germany in the 1950s or de Broglie in France, have tried to force physics in the direction of their own ideas; but where such mandarins succeed at all, it is only in one country, and only for a limited time. The direction of physics today is overwhelmingly set by young physicists, who are not yet weighed down with honors or authority, and whose influence—the excitement they stir up—derives from the objective progress that they are able to make. If our expression of the laws of nature is socially constructed, it is constructed in a society of scientists that evolves chiefly through grappling with nature's laws.

Some historians do not deny the reality of the laws of nature, but nevertheless refuse to take present scientific knowledge into account in describing the scientific work of the past.[25] This is partly to avoid anachronisms, like supposing that scientists of the past ought to have seen things in the way we do, and partly out

25. This point of view was expressed to me by the historian Harry Collins, then at the Science Studies Centre of the University of Bath.

of a preoccupation with maintaining the intellectual independence of historians.[26] The problem is that in ignoring present scientific knowledge, these historians give up clues to the past that cannot be obtained in any other way.

A historian of science who ignores our present scientific knowledge seems to me like a historian of U.S. military intelligence in the Civil War who tells the story of George McClellan's hesitations in the Virginia peninsula in the face of what McClellan thought were overwhelming Confederate forces without taking into account our present knowledge that McClellan was wrong. Even the choice of topics that attract the interest of historians has to be affected by what we now know were the paths that led to success. What Herbert Butterfield called the Whig interpretation of history is legitimate in the history of science in a way that it is not in the history of politics or culture, because science is cumulative, and permits definite judgments of success or failure.

Sokal was not the first to address these issues,[27] but he has done a great service in raising them so dramatically. They are not entirely academic issues, in any sense of the word "academic." If we think that scientific laws are flexible enough to be affected by the social setting of their discovery, then some may be tempted to press scientists to discover laws that are more proletarian or feminine or American or religious or Aryan or whatever else it is they want. This is a dangerous path, and more is at stake in the controversy over it than just the health of science. As I mentioned earlier, our civilization has been powerfully affected by the discovery that

26. In "Independence, Not Transcendence, for the Historian of Science," *ISIS* (March 1991), Paul Forman called for historians to exercise an independent judgment not just of how scientific progress is made, but even of what constitutes scientific progress.

27. See especially Gerald Holton, *Science and Anti-Science* (Harvard University Press, 1993), and Paul R. Gross and Norman Levitt, *Higher Superstition* (Johns Hopkins University Press, 1994). The issue of *Social Text* in which Sokal's hoax appeared was intended as a response to Gross and Levitt's book, which also, according to Sokal, inspired his hoax.

nature is strictly governed by impersonal laws. As an example I like to quote the remark of Hugh Trevor-Roper that one of the early effects of this discovery was to reduce the enthusiasm for burning witches. We will need to confirm and strengthen the vision of a rationally understandable world if we are to protect ourselves from the irrational tendencies that still beset humanity.

13

Science and Sokal's Hoax:
An Exchange

My article about Sokal's hoax evoked a good deal of comment, not unanimously favorable. Of the letters sent to *The New York Review of Books*, four were selected for publication a few months later, three of them containing serious criticisms. Michael Holquist, Professor of Comparative Literature at Yale, together with Robert Shulman, Yale's Sterling Professor of Molecular Biophysics and Biochemistry, had some nice things to say about my article, but took me to task for espousing a kind of dualism: on one hand science, on the other culture. Professor George Levine, of the Rutgers Center for the Critical Analysis of Contemporary Culture, objected to my statement that (with various qualifications) the conclusions of modern physics have no cultural implications. Professor M. Norton Wise, of the Princeton Program in the History of Science, criticized various aspects of what he called my ideology of science, and in particular found my comments on the work of Thomas Kuhn outrageous. *The New York Review* gave me the opportunity to respond to these letters in the same issue, and I answered with the following remarks.

I am grateful to those who sent comments on my article "Sokal's Hoax," including those who, by disagreeing with me, have given me this chance to take another whack at the issues it raised.

Professors Holquist and Shulman have me dead to rights in calling my views dualistic. I think that an essential element needed in the birth of modern science was the creation of a gap between the world of physical science and the world of human culture.[1] Endless trouble has been produced throughout history by the effort to draw moral or cultural lessons from discoveries of science. The physics and biology of Aristotle were largely based on a conception of naturalness, which was believed also to have moral and cultural implications, as for instance that some people are naturally slaves. After relativity theory became widely publicized in 1919, the Archbishop of Canterbury like many others conscientiously worried over the effect that relativity was going to have on theology, and had to be reassured by Einstein himself in 1921 that relativity had no implications for religion.[2] Professors Holquist and Shulman quote Emile Durkheim for the proposition that a gap between ways of viewing reality such as that between science and culture is characteristic of religious phenomena, but I think that just the opposite is true; if you want to find astronomy all muddled with cultural or moral values, you would turn to Dante's *Paradiso* rather than Galileo's *Dialogo*. In trying to revive a "holistic view of nature," Professors Holquist and Shulman are seeking to fill a much-needed gap.

Quantum mechanics provides a good example of the need to maintain this separation between physics and other forms of culture. Quantum mechanics has been variously cited as giving support to mysticism, or free will, or the decline of quantitative rationality. Now, I would agree that anyone is entitled to draw any inspiration he or she can from quantum mechanics, or from anything else. This is what I meant when I wrote that I had nothing to say against the use of science as metaphor. But there is a difference between inspiration and implication, and in talking of the "telling cultural implications" of quantum mechanics, Professor Levine may be confusing the two. There is simply no way that any cultural consequences can be *implied* by quantum mechanics. It

1. On this, see Herbert Butterfield in *The Origins of Modern Science* (Free Press, 1957), especially chapter 2.

2. Gerald Holton, *Einstein, History, and Other Passions* (Addison-Wesley, 1996), p. 129.

is true that quantum mechanics does "apply always and every-where," but what applies is not a proverb about diverse points of view but a precise mathematical formalism, which among other things tells us that the difference between the predictions of quan-tum mechanics and pre-quantum classical mechanics, which is so important for the behavior of atoms, becomes negligible at the scale of human affairs.

I suggest the following thought experiment. Suppose that physi-cists were to announce the discovery that, beneath the apparently quantum mechanical appearance of atoms, there lies a more fun-damental substructure of fields and particles that behave accord-ing to the rules of plain old classical mechanics. Would Professor Levine find it necessary to rethink his views about culture or phi-losophy? If so, why? If not, then in what sense can these views be said to be implied by quantum mechanics?

I was glad to see that Professor Wise, an expert on late-nine-teenth-century physics, finds no error in what I had to say about the history of science. Unfortunately he does find a great many errors in things that I did not say. I never said that there were no physicists in the early twentieth century who found cultural or philosophical implications in relativity or quantum mechanics, only that in my view these inferences were not valid. I never said that the apparent subjectivism of quantum mechanics was "of no great historical significance," only that I think we know better now. Just as anyone may get inspiration from scientific discoveries, scientists in their work may be inspired by virtually anything in their cultural background, but that does not make these cultural influences a permanent part of scientific theories. I never tried "to expunge all mystical physicists" as well as "creationists and other religious enthusiasts" from the history of science. I did say that I had never met a physicist who was a mystic, but my article had nothing to say about the frequency of other forms of religious be-lief among scientists, past or present.

On the subject of mystical physicists, it is interesting that when Professor Wise tries to find up-to-date examples, he can get no closer than two physicists whose major work was done more than sixty years ago. He expresses surprise that no physicist has yet presented string theory as a form of Platonic mysticism, but I think I can explain this. It is because we expect that string theory *will* be testable—if not directly by observing the string vibrations, then indirectly, by calculating whether string theory correctly accounts for all of the currently mysterious features of the Standard Model of elementary particles and General Relativity. If it were not for this expectation, string theory would not be worth bothering with.

I tried in my article to put my finger on precisely what divides me and many other scientists from cultural and historical relativists, by saying that the issue is not the belief in objective reality itself, but the belief in the reality of the laws of nature. Professor Wise makes a good point that, in judging the reality of the laws of nature, the test is not just their validity, but also their lack of "multiplicity." Indeed, as I wrote in my article, one of the things about laws of nature like Maxwell's equations that convinces me of their objective reality is the absence of a multiplicity of valid laws governing the same phenomena, with different laws of nature for different cultures.

(To be precise, I don't mean that there is no other valid way of looking at the electric and magnetic phenomena that Maxwell's equations describe, because there are mathematically equivalent ways of rewriting Maxwell's theory, and the theory itself can be replaced with a deeper theory, quantum electrodynamics, from which it can be derived. What I mean is that there is no valid alternative way of looking at the phenomena described by Maxwell's equations that does not have Maxwell's equations as a mathematical consequence.)

Whatever cultural influences went into the discovery of Maxwell's equations and other laws of nature have been refined away, like slag from ore. Maxwell's equations are now understood in the same way by everyone with a valid comprehension of electricity and magnetism. The cultural backgrounds of the scientists who discovered such theories have thus become irrelevant to the lessons that we should draw from the theories. Professor Wise and some others may be upset by such distinctions because they see them as a threat to their own "agenda," which is to emphasize the connections between scientific discoveries and their cultural context; but that is just the way the world is.

On the other hand, the gap between science and other forms of culture may be narrow or absent for the sciences that specifically deal with human affairs. This is one of the reasons that in writing of this gap in my article I wrote about physics, and explicitly excluded sciences like psychology from my remarks. I concentrated on physics also because that is what I know best. Professors Holquist and Shulman are mistaken in thinking that when talking of "science," I meant just physics, and excluded "microbiology, genetics, and the new brain sciences." I was pretty careful in my article to write of physics when I meant physics, and of science when I meant science. I can't see why Professor Shulman, a distinguished molecular biophysicist and biochemist, should be unhappy with my *not* offering opinions about the cultural implications of biology.

I should perhaps have made more clear in my article that I have no quarrel with most historians, philosophers, and sociologists of science. I am a fan of the history of science, and in my recent books I have acknowledged debts to writings of numerous historians, philosophers, and sociologists of science.[3] In contrast with Alan Sokal, who in perpetrating his hoax was mostly concerned about a breakdown of the alliance between science and the politi-

3. These include contemporary historians of science like Laurie Brown, Stephen Brush, Gerald Holton, Arthur Miller, Abraham Pais, and Sam Schweber; sociologists of science like Robert Merton, Sharon Traweek, and Stephen Woolgar; and philosophers like Mario Bunge, George Gale, Ernest Nagel, Robert Nozick, Karl Popper, Hilary Putnam, and W. V. Quine. These references can be found in *Dreams of a Final Theory* (Pantheon, 1993) and *The Quantum Theory of Fields* (Cambridge University Press, 1995). There are many others whose works I have found illuminating, including the historian of science Peter Galison, the sociologist of science Harriet Zuckerman, and the philosophers Susan Haack and Bernard Williams.

cal left, my concern was more with the corruption of history and sociology by postmodern and constructivist ideologies. Contrary to what Professor Levine may think, my opposition to these views is not due to any worry about the effects they may have on the economic pinch hurting science. In years of lobbying for federal support of scientific programs, I never heard anything remotely postmodern or constructivist from a member of Congress.

Among philosophers of science, Thomas Kuhn deserves special mention. He was a friend of mine whose writings I often found illuminating, but over the years I was occasionally a critic of his views.[4] Even in his celebrated early *Structure of Scientific Revolutions,* Kuhn doubted that "changes of paradigm carry scientists and those who learn from them closer and closer to the truth." I corresponded with him after we met for the last time at a ceremony in Padua in 1992, and I found that his skepticism had become more radical. He sent me a copy of a 1991 lecture,[5] in which he had written that "it's hard to imagine . . . what the phrase 'closer to the truth' can mean"; and "I am not suggesting, let me emphasize, that there is a reality which science fails to get at. My point is rather that no sense can made of the notion of reality as it has ordinarily functioned in philosophy of science." I don't think that it was "outrageous" for me to have said that, as I understood his views, Kuhn denied the objective nature of scientific knowledge.

Professor Levine and several others object to my criticism of Jacques Derrida, based as it seems to them on a single paragraph chosen by Sokal for mockery, which begins, "The Einsteinian constant is not a constant, is not a center. It is the very concept of variability—it is, finally, the concept of the game." When, in reading Sokal's *Social Text* article, I first encountered this paragraph, I was

4. See *Dreams of a Final Theory,* and "Night Thoughts of a Quantum Physicist," Essay 9 above.

5. Thomas Kuhn, "The Trouble with the Historical Philosophy of Science," Rothschild Distinguished Lecture, November 19, 1991 (Department of the History of Science, Harvard College, 1992).

bothered not so much by the obscurity of Derrida's terms "center" and "game." I was willing to suppose that these were terms of art, explained elsewhere by Derrida. What bothered me was his phrase "the Einsteinian constant," with which I had never met in my work as a physicist. True, there is something called Newton's constant which appears in Einstein's theory of gravitation, and I would not object if Derrida wanted to call it "the Einsteinian constant," but this constant is just a number (0.00000006673 in conventional units), and I did not see how it could be the "center" of anything, much less the concept of a game.

So I turned for enlightenment to the talk by Derrida from which Sokal took this paragraph. In it, Derrida explains the word "center" as follows: "Nevertheless, . . . structure—or rather, the structurality of structure—although it has always been involved, has always been neutralized or reduced, and this by a process of giving it a center or referring it to a point of presence, a fixed origin."[6] This was not much help.

Lest the reader think that I am quoting out of context, or perhaps just being obtuse, I will point out that, in the discussion following Derrida's lecture, the first question was by Jean Hyppolite, professor at the Collège de France, who, after having sat through Derrida's talk, had to ask Derrida to explain what he meant by a "center." The paragraph quoted by Sokal was Derrida's answer. It was Hyppolite who introduced "the Einsteinian constant" into the discussion, but while poor Hyppolite was willing to admit that he did not understand what Derrida meant by a center, Derrida just started talking about the Einsteinian constant, without letting on that (as seems evident) he had no idea of what Hyppolite was talking about. It seems to me that Derrida in context is even worse than Derrida out of context.

6. Jacques Derrida, "Structure, Sign, and Play in the Discourse of the Human Sciences" in *The Structuralist Controversy*, ed. R. Macksey and E. Donate (Johns Hopkins University Press, 1972), p. 247.

14

Before the Big Bang

This is a review that Robert Silvers of *The New York Review of Books* asked me to do of three 1997 books on cosmology: Tim Ferris's *The Whole Shebang: A State-of-the Universe(s) Report* (Simon and Schuster); Alan Guth's *The Inflationary Universe: The Quest for a New Theory of Cosmic Origins* (Helix Books/Addison-Wesley); and Martin Rees's *Before the Beginning: Our Universe and Others* (Simon and Schuster, published in the United States by Addison-Wesley). The review turned out to be mostly a straightforward nontechnical account of some current issues in cosmology, but I also suggested some lessons that research in cosmology may have for the history and philosophy of science.

Astronomical measurements since the time that this article was written have not challenged the main features of the Big Bang theory described in this review, but they have clarified some very important features of the theory. These are described in an "update," written for this collection, which follows the original review.

On a summer weekend a few years ago my wife and I visited friends at their ranch in the Glass Mountains of West Texas. After dinner we sat outside on lawn chairs and looked up at the sky. Far from city lights, in clear, dry air, and with the moon down, we could see not only Altair and Vega and the other bright stars that you can see from anywhere on a cloudless summer night, but also an irregular swath of light running across the sky, the Milky Way, as I had not seen it in decades of living in cities.

The view is something of an illusion: the Milky Way is not

something out there, far from us—rather, we are in it. It is our galaxy: a flat disk of about a hundred billion stars, almost a hundred thousand light years across, within which our own solar system is orbiting, two thirds of the way out from the center. What we see in the sky as the Milky Way is the combined light of the many stars that are in our line of sight when we look out along the plane of the disk, almost all of them too far away to be seen separately. Staring at the Milky Way and not being able to make out individual stars in it gave me a chilling sense of how big it is, and I found myself holding on tightly to the arms of my lawn chair.

Astronomers used to think that our galaxy was the whole universe, but now we know that it is one of many billions of galaxies, extending out billions of light years in all directions. The universe is expanding: any typical galaxy is rushing away from any other with a speed proportional to the distance between them. As the universe becomes less crowded it is also becoming colder. Observing what is happening now, and using what we know of the laws of physics, we can reconstruct what must have been happening in the past.

Here is the account that is now accepted by almost all working cosmologists. About 10 to 15 billion years ago, the contents of the universe were so crowded together that there could be no galaxies or stars or even atoms or atomic nuclei. There were only particles of matter and antimatter and light, uniformly filling all space. No definite starting temperature is known, but our calculations tell us that the contents of the universe must once have had a temperature of at least a thousand trillion degrees Centigrade. At such temperatures, particles of matter and antimatter were continually converting into particles of light, and being created again from light. Meanwhile, the particles were also rapidly rushing apart, just as the galaxies are now. This expansion caused a fast cooling of the particles, in the same way that the expansion of freon gas in a refrigerator's coils cools the appliance. After a few seconds, the temperature of the matter, antimatter, and light had dropped to about ten billion degrees. Light no longer had enough energy to turn into matter and antimatter. Almost all matter and antimatter particles annihilated each other, but (for reasons that are somewhat mysterious) there was a slight excess of matter particles—electrons, protons, and neutrons—which could find no antimatter particles to annihilate them, and they therefore survived this great extinction. After three more minutes of expansion the leftover matter became cold enough (about a billion degrees) for protons and neutrons to bind together into the nuclei of the lightest elements: hydrogen, helium, and lithium.

For three hundred thousand years the expanding matter and light remained too hot for nuclei and electrons to join together as atoms. Stars or galaxies could not begin to form because light exerts strong pressure on free electrons, so any clump of electrons and nuclei would have been blasted apart by light pressure before its gravity could begin to attract more matter. Then, when the temperature dropped to about three thousand degrees, almost all electrons and nuclei became bound into atoms, in what astronomers call the epoch of recombination. (The "re" in "recombination" is misleading. At the time of recombination electrons and nuclei had never before been combined into atoms.) After recombination, gravitation began to draw matter together into galaxies and then into stars. There it was cooked into all the heavier elements, including those like iron and oxygen from which, billions of years later, our earth was formed.

This account is what is commonly known as the Big Bang cosmology. As the term is used by cosmologists, the Big Bang was not an explosion that occurred sometime in the past; it is an explosion involving all of the universe we can see, that has been going on for 10 to 15 billion years, since as far back in time as we can reliably trace the history of the universe; and it will doubtless continue for billions of years to come, and perhaps forever.

Though the Big Bang cosmology is a great scientific achievement, it has not answered the most interesting question: What is the origin of the universe?—not what happened in the first three hundred thousand years of the Big Bang, or in its first three minutes, or in its first few seconds, but at the very start, if there was one, or even before. In 1992 I was among a group of physicists trying to sell Vice President Al Gore on the need for a large new elementary particle accelerator, the Superconducting Super Collider. Knowing that even people who couldn't care less about the laws of physics are curious about cosmology, we described how this instrument might lead to the discovery of mysterious particles that are thought to inhabit the voids between the galaxies and to make up most of the matter of the universe. The Vice President listened politely and promised warm support from the Clinton administration.[1] Then, just as he was leaving, he turned back into the room, and diffidently asked if we could tell him what happened before the Big Bang.

I don't recall what we answered, but I am sure that it was not very enlightening. No one is certain what happened before the Big Bang, or even if the question has any meaning. When they thought about it at all, most physicists and astronomers supposed until recently that the universe started in an instant of infinite temperature and density at which time itself began, so that questions about what happened before the Big Bang are meaningless, like questions about what happens at temperatures below absolute zero. Some theologians welcome this view, presumably because it bears a resemblance to scriptural accounts of creation. Moses Maimonides taught that "the foundation of our faith is the belief that God created the Universe from nothing; that time did not exist previously, but was created."[2] Saint Augustine thought the same.

But opinions among cosmologists have been shifting lately, toward a more complicated and far-reaching picture of the origin of the universe. This new view is given prominence in the books un-

1. That support turned out to be tepid, and the project was canceled by Congress.

2. This is taken from the second revised edition of the M. Friedlander translation of *The Guide for the Perplexed* (Routledge and Kegan Paul, 1904; reprinted by Dover Publications, 1956), p. 212.

der review by Timothy Ferris, Alan Guth, and Sir Martin Rees. All three books also give clear introductions to the standard Big Bang theory and to the physical theories used by cosmologists.

Ferris is an emeritus professor of journalism at the University of California at Berkeley, and an experienced writer on science for the general reader. He has an attractive, cool style, and his book is broad in scope, extending even into philosophical matters, but without the gushing or silliness that sometimes marks discussions of cosmology. Although he is not a working scientist, Ferris has a sure grasp of the science he describes. I do not think there is a better popular treatment of some of the topics that Ferris covers. (He tells some good stories too. Since the late 1960s, he writes, the physicist Ray Davis has been measuring the rate at which the sun emits the elusive particles known as neutrinos, by catching one now and then in a large tank of perchlorethylene, a common dry-cleaning fluid. After ordering 100,000 gallons of perchlorethylene from a chemical supply house, Davis began to receive advertisements from a supplier of coat hangers.)

Rees is an eminent astrophysicist, currently the fifteenth Astronomer Royal of England and professor at the University of Cambridge, and one of the best expositors of his subject. Because he has made important contributions to many problems of astrophysics, Rees is able to give the reader a better sense of how science is done than most philosophers of science, in the way that a good military memoir—by Ulysses S. Grant, say, or Omar Bradley—gives a better sense of the nature of war than the generalities of Sun Tzu or Carl von Clausewitz. He makes some telling points about science policy: he is, for example, dead right about the harm done to astronomical research by our commitment to the manned space program. He quotes Riccardo Giacconi, the first director of the Space Telescope Science Institute, as saying that if from the beginning it had been planned that an unmanned rocket instead of the space shuttle would put the Hubble Space Telescope into orbit, then seven similar telescopes could have been built and launched for what so far has been spent on just one. Rees's book also evocatively describes scientists he has known—not only those, like Stephen Hawking and Roger Penrose, whose names are familiar to the public, but others who have also made major contributions to astrophysics, including Subrahmayan Chandrasekhar, Robert Dicke, John Wheeler, Yakov Zeldovich, and Fritz Zwicky.

Guth is a young professor of physics at MIT. Though this is his first book for the general reader, Guth writes engagingly and understandably. His book concentrates on research on the beginning of the universe, to which he has made a major contribution, and so it is an important historical document as well as a pleasure to read.

In December 1979 Guth was a postdoctoral research associate visiting the Stanford Linear Accelerator Center, worrying about where he would get his next job. Together with a Cornell colleague, Henry Tye, he was studying the cosmological effects of certain physical fields. Fields are conditions of space itself, considered apart from any matter that may be in it. Fields can change from moment to moment and from point to point in space, in something like the way that temperature and wind velocity are conditions of the air that can vary with time and position in the atmosphere. The most familiar example of a field is the gravitational field, which we all feel tugging us toward the center of the earth. Most people have also felt the magnetic field pulling a piece of iron held in the hand toward the north or south pole of a bar magnet. In the modern theory of elementary particles known as the Standard Model, a theory that has been well verified experimentally, the fundamental components of nature are a few dozen different kinds of field.

The fields considered by Guth in this work are called scalar fields, which means that they are made up of quantities that like the air temperature are purely numerical, in contrast to gravitational and magnetic fields, which like wind velocity point in a definite direction. Scalar fields do not tug at anything, so we are not normally conscious of them, but physicists think that they pervade the present universe. In the simplest version of the Standard Model of elementary particles, it is the action of scalar fields on electrons and quarks and other elementary particles that gives these particles their masses.

All fields can carry energy, so these scalar fields can give an energy even to otherwise empty space. According to Einstein's General Theory of Relativity, all forms of energy affect the rate of expansion of the universe. To judge from the measured rate of expansion of the universe, empty space now has almost no energy. But the energy of any field naturally depends on its strength. For instance, no one notices the energy in the weak magnetic field of an ordinary bar magnet, but the much stronger magnetic fields of modern electromagnets have energies that can physically wreck the magnets if they arc not carefully designed. The strengths of the scalar fields considered by Guth were different under the conditions of the early universe from what they are now; they gave "empty space" an enormous energy, quite unlike the zero-energy space we live in now.

Guth calculated that in the early universe the energy of the scalar fields would have remained constant for a while as the universe expanded, which would produce a constant rate of expansion, in contrast to the situation in the present universe, where the rate of expansion decreases as the density of matter decreases. With a constant rate of expansion the universe would have grown exponentially, like a bank account with a constant rate of compound interest, but with the size of the universe doubling again and again in each tiny fraction of a second. Guth called this phenomenon "inflation." The possibility of an exponential expansion had been realized by others, including Andrei Linde and Gennady Chibisov at the Lebedev Physical Institute in Moscow, and in itself this would have been a technical result of interest only to other physicists. But then it occurred to Guth that the existence of an era of inflation would solve one of the outstanding problems of cosmology, known as the "flatness problem."

The problem is to understand why the curvature of space was so small in the early universe. General Relativity tells us that space can be curved, and that this curvature has an effect on the rate of expansion of the universe that is similar to the effect of the energy in matter and in the scalar fields. Other things being equal, the greater the curvature the faster the expansion. We don't know precisely what the curvature of space is right now, but from measurements of the rate of expansion and the amount of matter in clusters of galaxies we do know that the energy of matter accounts for at least 10 percent of the rate of expansion, and maybe all of it. This leaves at most 90 percent of the expansion rate to be produced by curvature. But as the universe has been expanding, the density of matter has been falling, so the fraction of the expansion rate due to curvature has been steadily growing—if it is no more than 90 percent now, then in the first second of the Big Bang it must have been less than about one part in a thousand trillion. This is not a paradox—there is no reason why the curvature should not have been very small—but it is the sort of thing physicists would like to explain if they could.

What Guth realized was that during inflation the fraction of the expansion rate caused by curvature would have been rapidly decreasing. (The reason is that the curvature was itself decreasing, like the surface curvature of an expanding balloon, while the energy of the scalar fields remained roughly constant.) So to understand why space was so flat at the beginning of the present Big Bang it is not necessary to make any arbitrary assumptions; if the Big Bang was preceded by a sufficient period of inflation, it would necessarily have started with negligible curvature. Guth wrote in his diary that inflation "can explain why the universe [was] so incredibly flat," and added a heading, "SPECTACULAR REALIZATION." So it was.

Guth soon also discovered that inflation would solve other cosmological puzzles, some of which he had not even realized were puzzles. One of these is known as the "horizon problem." Conditions in the universe at the time of recombination—when almost all electrons and nuclei became bound into atoms—seem to have been pretty uniform over distances of at least ninety million light years. This is revealed by observation of microwave radiation left over from that time. The problem is that at recombination there had not yet been enough time since the beginning of the Big Bang for light or anything else to have traveled more than a small fraction of this distance, so that no physical influence would have been able to stir things up to form the uniform texture we observe.

Like the flatness problem, this is not a logical contradiction—there is no reason why the universe could not have started out perfectly uniform—but, again, it is the sort of thing we hope to explain. Guth found that inflation provides an explanation: during the inflationary era the part of the universe that we can observe would have occupied a tiny space, and there would have been plenty of time for everything in this space to be homogenized. Guth gave a talk about his work a few weeks later, and the next day received two job offers and three invitations to give talks at other physics departments.

Then Guth did something that might seem odd: he went to work to find things wrong with his inflation theory. This is the way physics works; Guth must have known that if something was wrong with inflationary cosmology then this would soon become clear to many physicists, and Guth would naturally want to be the first one to find that something was wrong with his theory and to have a chance to fix it. It is thus that the existence of a common standard of judgment leads physicists, who are no more saintly than economists, to question their own best work.

As it turned out, there *was* something wrong with the original version of inflation theory. Andrei Linde in Moscow seems to have been the first to realize this, but it was discovered independently by a host of others: Guth and Erick Weinberg of Columbia University; Paul Steinhardt and Andreas Albrecht at the University of Pennsylvania; and Stephen Hawking, Ian Moss, and John Stewart at the University of Cambridge. The difficulty had to do with the end of inflation. Guth had originally assumed it ended with what is called a phase transition, like the freezing of water at zero degrees Centigrade or its boiling at 100 degrees.

The phase transition at the end of inflation was of course different from the one in which water turns into ice or vapor. When the temperature of the universe dropped to about a million trillion trillion degrees, the scalar fields jumped from their original values—the numbers that characterize the strengths of the fields—to the values they have now. During this transition bubbles of ordinary zero-energy empty space would have formed here and there, like bubbles of vapor formed by boiling water. The energy that had been in the scalar fields during inflation would have wound up on the surfaces of the bubbles.

Guth had thought at first that these bubbles would have merged and disappeared, spreading the energy on their surfaces evenly throughout space, after which the universe could be described by the conventional Big Bang theory. But calculations showed that although the bubbles were expanding rapidly, the universe was expanding faster, so that the bubbles never could have merged with one another. These calculations posed an immediate difficulty for Guth: there would be no place in such a universe that would look like the Big Bang in which we find ourselves. Since we now live in zero-energy space, our own Big Bang could only be located inside one of these bubbles; but by the end of inflation the bubbles would have expanded so much that the density of matter and light inside them would be infinitesimal.

Linde and—independently—Albrecht and Steinhardt also found a way out of this difficulty. Making different assumptions about the physical forces responsible for the release of the space energy, they showed that this energy could have leaked into the interior of the bubbles of ordinary space during inflation, eventually turning into matter and light at temperatures around a million trillion trillion degrees. This matter and light would then have expanded and cooled as in the conventional Big Bang theory. The part of the universe we observe, extending billions of light years and containing billions of galaxies, thus would be just a tiny part of the interior of one of these bubbles. There would be countless other bubbles of ordinary space, too far away to see, and presumably many of these would develop into big bangs like our own. Soon after this work,

Guth gave a talk at the Harvard-MIT Joint Theoretical Seminar, with the subtitle "How Linde and Steinhardt Solved the Problems of Cosmology, While I Was Asleep."

Loosely put, each bubble of ordinary space could be called a "universe"—this is what Ferris and Rees mean when they refer to "universe(s)" and "our universe and others" in their subtitles. If instead we stick to the usual definition of "universe" as everything, the whole shebang, then the idea of a multiplicity of big bangs, if correct, would represent an enormous enlargement of our notion of the universe. It would be the third step in a historical progression that started with the suggestion in 1584 by Giordano Bruno that the stars are suns like our own, and continued with the demonstration by Edwin Hubble in 1923 that many faint patches of light in the sky are galaxies like our own.

This "new inflationary cosmology" has its own internal problems. Since 1980 other inflation theories have proliferated. To me, the most interesting is the "chaotic inflation" theory of Linde (now at Stanford). He makes the reasonable assumption that the scalar fields did not start at the beginning of time with some definite value, uniform throughout the universe, but instead were fluctuating wildly, so that inflation began here and there at different times.

Chaotic inflation opens up the possibility I mentioned earlier, of a new view of what happened before our Big Bang. If the scalar fields don't evolve in lock step everywhere in the universe, then very far away there may have been other big bangs before our own, and there may be others yet to come. Meanwhile the whole universe goes on expanding, so there is always plenty of room for more big bangs. Thus although our own Big Bang had a definite beginning about ten to fifteen billion years ago, the bubbling up of new big bangs may have been going on forever in a universe that is infinitely old.

An even stranger idea has been gaining ground lately. Just as what we have been calling the universe may be only a tiny part of the whole, so also what we usually call the "constants of nature," like the masses we ascribe to the elementary particles, may vary from one part of the universe to another. Inflationary cosmology offers a concrete realization of this idea. The evolution of the scalar fields within each expanding bubble may lead to final values of these fields that differ from one bubble to another, in which case each big bang would wind up with different values for physical constants. (The Harvard theorist Sidney Coleman has shown how something like this can happen even apart from inflation when quantum mechanics is applied to the whole universe.) In any case, if for some reason or other the constants of nature vary from one part of the universe to another, then it would be no mystery why these constants are observed to have values that allow for the appearance of intelligent life: why, for instance, the charge and mass of the electron are so small. Because the charge and mass are small, the force between electrons and quarks is too weak to keep electrons inside atomic nuclei, so that electrons in atoms form clouds outside the nuclei, which hold atoms together in chemical molecules, including those necessary for life. Only in the parts of the universe where the constants have such values is there anyone to worry about it.

Arguments of this sort may explain why in the present universe there is almost no energy in empty space, despite the fact that according to quantum mechanics there are continual fluctuations in the gravitational and electromagnetic and other fields, fluctuations that by themselves would give empty space an enormous energy. Here and there in the universe there are regions where by chance the scalar fields happen to wind up at the end of inflation with negative energies that cancel almost all the energy of the field fluctuations. In the far more numerous regions where empty space winds up with a large amount of energy, there are forces that would prevent the formation of stars and galaxies; and so there would, in those regions, be no one who could raise the question of the energy of space.

This sort of reasoning is called anthropic, and it has a bad name among physicists. Although I have used such arguments myself in some of my own work on the problem of the vacuum energy,[3] I am not that fond of anthropic reasoning. I would personally be much happier if we could precisely calculate the values of all the constants of nature on the basis of fundamental principles, rather than having to think about what values are favorable to life. But nature cares little about what physicists prefer.

How much of all this are we to believe? As Ferris, Guth, and Rees all make clear, in answering this question one must distinguish between the big Bang theory itself, which describes what happened once the temperature of the observable part of the universe dropped below a few trillion degrees, and the inflationary cosmologies, which try to account for what happened earlier.

About the Big Bang theory we can be quite confident. Our understanding of the laws of physics is good enough to allow us to trace the history of the Big Bang back to a time when the temperature of the universe was a thousand trillion degrees. Also, until the formation of galaxies, conditions in the universe were much the same everywhere, so that in our calculations we don't have to deal with complicated differences among local conditions of the sort that we find here on earth, which make it so difficult to predict whether it will rain next week.

The Big Bang theory is also confirmed by the discovery of various relics of the early universe. The most dramatic such relic is a whisper of microwave radiation that was produced in the epoch of recombination, radiation that has been cooled by the thousand-fold expansion of the universe since then to a temperature of 2.73 degrees above absolute zero.

The most convincing quantitative evidence for the Big Bang theory comes from another relic: five isotopes of the lightest elements, found spectroscopically in interstellar matter that has not yet been processed into stars. The measured abundances of these five isotopes generally agree nicely with calculations of the amounts of the same isotopes which, according to the theory, were produced by nuclear reactions at the end of the first three minutes.

---

3. In part with Hugo Martel and Paul Shapiro, of the University of Texas.

The Big Bang theory is not a temporary theoretical fashion, likely to be blown away by the next round of astronomical observation, but is almost certain to endure as part of any future theory of the universe. Ferris remarks that this conclusion "may seem curious to readers of the many newspaper and magazine articles that have appeared during the past decade proclaiming that this or that observational finding has put the big-bang theory in jeopardy." He quotes *Time* magazine reporting that the Big Bang theory is "unravelling." Journalists generally have no bias toward one cosmological theory or another, but many have a natural preference for excitement. It is exciting to report that some new observation threatens to throw the Big Bang theory into the dustbin of history. It is dull to report that although some detail of the theory has been put in question, the Big Bang theory itself is doing well. It is like one of those dull headlines that journalism students are warned about, such as "Crime Rates Remain Low in Toronto."

To be fair, I should add that overheated science journalism is occasionally abetted by scientists. Martin Rees has hard words for some astronomers, and justly points out that "journalists sometimes need to assess scientists' claims with as much skepticism as they customarily bring to those of politicians."

Although the Big Bang theory is overwhelmingly the consensus view of physicists and astronomers, you can still find dissidents among respected scientists who have a longstanding stake in other theories. One alternative is the "steady state" theory, in which there is no evolution of the universe as a whole; rather, new matter is always being created to fill up the gaps between the receding galaxies. Sir Fred Hoyle, one of the authors of the steady state theory, coined the term "big Bang theory" in order to poke fun at the consensus. The idea that the universe had no start appeals to many physicists philosophically, because it avoids a supernatural act of creation. And the idea that the universe does not evolve is attractive pragmatically, because if it is true at all, it could only be true under stringent conditions on the contents of the universe, and these conditions would give us an extra handle on the problem of explaining why things are the way they are. Chaotic inflation has in a sense revived the idea of a steady state theory in a grander form; our own Big Bang may be just one episode in a much larger universe that on average never changes. But the original form of the steady state theory as an alternative to the Big Bang was convincingly ruled out by the discovery of the 2.73-degree cosmic microwave radiation.

It is conceivable that some of the skeptics will turn out to be right about the Big Bang theory, but this seems unlikely. Rees cautiously gives odds of only 10 to 1 in favor of the Big Bang, but he quotes Yakov Zeldovich as saying that the Big Bang is "as certain as that the Earth goes round the Sun." At least within the past century, no other major theory that became the consensus view of physicists or astronomers—in the way that the Big Bang theory has—has ever turned out to be simply wrong. Our theories have often turned out to be valid only in a more limited context than we had thought, or valid for reasons that we had not understood. But they are not simply wrong—not in the way, for instance, that the cosmology of Ptolemy or Dante is wrong. Consensus is forced on us by nature itself, not by some orthodox scientific establishment; as Rees says, "Philosophers of science would be surprised at how many astronomers are eager rather than reluctant to join a revolutionary bandwagon."

Among skeptics outside the sciences, there are those multiculturalists who don't so much disagree with the standard cosmological theory as avoid the question of its objective truth. They see modern science as an expression of our "Western" civilization; it works for us, but the belief that the Milky Way is a river in the sky worked for the Mayans, and the belief that the Milky Way is a great canoe rowed by a one-legged paddler worked for the early peoples of the Amazon basin, so who can say that one belief is better than another?

I can. For one thing, modern cosmology is not confined within Western culture. European astronomy received important contributions from Egypt, Babylon, Persia, and the Arabs, and today astronomy and physics are pursued in the same way throughout the world. The West is not so unanimous about science; it has no shortage of believers in astrology, in "Heaven's Gate," and similar nonsense.

Apart from its inaccuracy, there is a certain risk in the attempt to tie modern cosmology to Western civilization. Whether or not the Mayans and early Amazonians and other ancient peoples believed in the objective truth of their theories of the Milky Way—they may have just used these theories to suggest convenient markers in the sky, like our own Big Dipper—they certainly did not know that the Milky Way is actually a disk containing billions of stars like our sun, much like billions of other galaxies throughout the universe. Anyone who became convinced that modern cosmology was peculiarly Western, and who did care about objective truth, might reasonably conclude that Western civilization is superior to all others in at least one respect, that in trying to understand what you can see in the sky on a starry night, Western astronomers got it right.

In contrast to the Big Bang theory, the theory of inflation is a good idea that explains a lot, but we can't yet be confident that it is right. There is no consensus here. Rees quotes Roger Penrose as saying that inflation is a "fashion the high-energy physicists have visited on the cosmologists" and that "even aardvarks think their offspring are beautiful." I don't agree with Penrose, but it is certainly true that it will be difficult to settle on a specific inflationary cosmology and decide if it is correct, for reasons both of astronomy and of physics.

The only relic we know that would have survived from the era of inflation and that would allow a quantitative astronomical test of the theory is the pattern of nonuniformities in matter and light. Quantum mechanics tells us that during inflation there must have been small fluctuations in the scalar fields. At recombination three hundred thousand years later, these fluctuations would show up as tiny nonuniformities in the temperature of matter and light. Those nonuniformities could be observed in the microwave radiation that comes to us from that time. So far, observations from the Cosmic Background Explorer satellite have given results that agree with the predictions of inflationary cosmology, but the observed nonuniformity is pretty much what had been widely expected before anyone thought of inflationary cosmology.

A critical test of these new theories will have to wait until new microwave telescopes can study finer details in the cosmic microwave radiation. Even then, it may not be possible to decide for or against inflation, both because these observations are clouded by radio noise from our own galaxy and because by now there are so many versions of inflationary cosmology. As we make progress in understanding the expanding universe, the problem itself expands, so that the solution seems always to recede from us.

I doubt if it will be possible to decide which version if any of inflationary cosmology is correct on the basis of astronomical observation alone, without fundamental advances in physics. Our present theories of elementary particles are only approximations, which don't adequately describe conditions at the time of inflation. We don't even know for sure whether the scalar fields really exist or, if they do, what different types there may be. This question will be settled at least in part by experiments at the next big accelerator, the Large Hadron Collider under construction near Geneva.

We shall also have to solve the old problem of reconciling the theory of gravitation and the principles of quantum mechanics. Rees makes a good point that so far we have generally been able to dodge this problem because gravitation and quantum mechanics are almost never both important in the same context. Gravitation governs the motions of planets and stars, but it is too weak to matter much in atoms, while quantum mechanics, though essential in understanding the behavior of electrons in atoms, has negligible effects on the motions of stars or planets. It is only in the very early universe that gravitation and quantum mechanics were both important. Famous theorems of Roger Penrose and Stephen Hawking use General Relativity to show that there must have been a definite beginning to the universe, but their proofs do not take quantum mechanics into account and are therefore inconclusive.

Our best hope for a quantum theory of gravitation lies in the speculative class of theories known as superstring theories.[4] Ferris says eloquently that superstring theory has become "a repository of the highest hopes of the finest minds in physics." Unfortunately, these theories have not yet settled down to a final form, and so far are not confirmed by the success of any quantitative predictions. It may be a long time before we can use these theories to decide whether the universe had a definite beginning, or whether the bubbling up of new big bangs has been going on forever.

About one thing I am sure. Those who think that an infinitely old universe is absurd, so there must have been a first moment in time, and those who think that a first moment is absurd, so the universe must be infinitely old, have one thing in common: whichever side happens to be right about the origin of the universe, the reasoning of both sides is wrong. We don't know if the universe is infinitely old or if there was a first moment; but neither view is absurd, and the choice between them will not be made by intuition, or by philosophy or theology, but by the ordinary methods of science.

2001 Update

This article needs updating in two important respects. I indicated in it that the energy density in empty space is very small or zero, and that the curvature of space is not responsible for more 90 percent of the present rate of expansion of the universe. Since this article was written astronomers using optical telescopes have used

4. In superstring theories the various elementary particles are explained as various modes of vibration of tiny one-dimensional rips in spacetime known as strings. This seems to run counter to the lesson of quantum field theory, that the different particles are bundles of the energy and momentum in different types of field. But at the relatively low energies that can be probed in today's accelerator laboratories, any superstring theory will look like a field theory.

observations of supernovas to measure the distances of galaxies that are rushing away from us at high speeds, while others using radio telescopes have measured fine details in the angular distribution of the cosmic microwave radiation left over from the early universe. From these measurements it can be inferred with pretty high confidence that the curvature of space is quite small, contributing much less than half the rate of expansion of the universe, which is what would be expected from typical theories of cosmological inflation. It can also be inferred that the energy in "empty" space, although vastly less than what would be expected from fluctuations in quantum fields, is nevertheless not zero, but rather contributes about 70 percent of the rate of expansion of the universe. This means that there are now *two* problems surrounding the energy of open space (also known as the cosmological constant.) One is the old problem why the energy of open space is so much smaller than would be expected on the basis of quantum field theory. The other is why its actual value is of the same order of magnitude as the energy density that is needed to account for the rate of expansion of the universe *at the present moment,* rather than in the distant past or future.

15

Zionism and Its Adversaries

The First Zionist Congress met in Basel in 1897, under the leadership of Theodor Herzl. Though Jews had never entirely left Palestine, this Congress is usually taken as the beginning of their modern effort to resettle in their ancestral home. In 1997 Martin Peretz, the editor in chief of *The New Republic*, decided to commemorate the one hundredth anniversary of the First Zionist Congress with a special double issue. Peretz possibly knew that I was concerned about the future of Israel because I had asked him for help in raising funds for The Jerusalem Winter School of Theoretical Physics, which I founded in 1983. Whether for this reason or some other, I was one of the sixteen people asked to contribute a short article.

My article focused on the surprising opposition to Zionism among liberal Westerners. Of course Zionism and Israel face more dangerous enemies—Arab irredentists, Muslim zealots, and ordinary old-fashioned anti-Semites. I concentrated on liberal Westerners because that is what I am, and for me there is a special pain in seeing some of my fellow Western liberals hostile to an ideal and a country I admire. This problem is still with us, as shown for instance by the October 2000 condemnation of Israel by the United Nations Security Council for responding to violence initiated by Palestinian Arabs, violence that broke out after far-reaching concessions by the government of Israel.

This article has nothing to say about physics or cosmology. Nevertheless, it seemed to me that its particular slant made it appropriate for inclusion in this collection. As I say in the article, one of the things that makes me sympathetic to Zionism is that it represents the intrusion of a democratic, scientifically sophisticated, secular culture into a part of the world that for centuries had been despotic, technically backward, and obsessed with religion. The values of science and secularism that are disliked by some liberal intellectuals in the West are the same ones that I have tried to defend throughout this collection.

I write about Zionism as one who has no interest in the preservation of Judaism (or, I hasten to add, any other religion), but a great deal of interest in the preservation of Jews. There always were two different ways that Zionists viewed the return to Zion: as a duty for all Jews, imposed by their religion, or as an opportunity for those Jews who want or need to live in a Jewish nation. As an unreligious American Jew, I feel no desire or duty to change my nationality, and I am in no position to deplore the fact that few other American Jews want to become Israelis. But I very much hope that the land of Israel will continue to provide an opportunity for Jews to flee oppression in any country, as it has since 1948. This refuge was needed in the twentieth century far more than could ever have been expected in 1897, but because of Arab and British opposition it was tragically not available at the time of greatest need, in the 1930s and 1940s. Zionism today in part reflects the determination that Israel will be available as a refuge the next time it is needed.

Zionism also represents the intrusion—by purchase and settlement rather than conquest, at least until Arab assaults made military action necessary—of a democratic, scientifically sophisticated, secular culture into a part of the world that for centuries had been despotic, technically backward, and obsessed with religion. For me, it is this essentially Western character of Zionism that gives it an attraction that goes beyond its defensive role. It will be betrayed if Orthodox zealots succeed in making Israel a theocratic state, but I can't believe that this will happen.

It is in part the Western character of Zionism that makes it so hateful to many others—not just to Muslims, but also to Westerners who for one reason or another find it comfortable to line up with the adversaries of Western civilization. It is very convenient by attacking Zionism to express solidarity with the poorer non-Western people of world, without having actually to make sacrifices to help them. In this way, anti-Zionism has come to play the role of a gutter multiculturalism.

Of course, this is not the whole story. Anti-Zionism also serves to give vent to deep-seated anti-Semitic feelings, and it helps to appease Muslim sentiment and thus to maintain access to Middle Eastern oil. Out of this tangle of motivations came the hysterical anti-Zionism of the last decade of Stalinism; the resolution by a 1975 world congress on women held at Mexico City a few years ago, that Zionism is oppressive to women; the infamous resolution of the General Assembly of the United Nations that same year equating Zionism with racism; and the continuing one-sidedness of international opinion regarding the relations between Israel and the Palestinian Arabs.

As a liberal I especially deplore a peculiar even-handedness on the part of many fellow liberals, an even-handedness that sees no moral difference between the efforts of Israel to preserve its existence in a fraction of a fraction of the original Palestine mandate and the attacks of those who wish to destroy it, and that views the building of housing projects in Jerusalem as being on a moral plane with the firing of machine guns at school buses. Do I need to say that there is in fact a great moral distinction between the democratic, secular Israel created by Zionism, whose long-range aim is simply to be left in peace, and the enemies that surround it? It is a distinction that ought to make liberals see Zionism again as they used to, as a natural part of the liberal agenda.

16

The Red Camaro

The magazine *George* used to have an "Eyewitness" feature, consisting of a brief account of some important moment that had occurred just thirty years earlier. Somehow its editors found out in 1997 that my first paper on the eletroweak theory was written in 1967, so they asked me to contribute a short article describing just how I came to do this work. *George* was a glossy magazine filled with fashion advertisements, the sort of magazine to which I would never have thought of contributing, but they offered a fee per word higher than any I had ever received, so I suppressed my puritan instincts and wrote the piece below. The editorial staff at *George* was very helpful, and we wound up with the fullest account I have given of the circumstances surrounding my work on the electroweak theory.

As it turned out, the article appeared in the magazine on a page opposite an advertisement for Liz Claiborne perfume. The ad was printed in scented ink, so anyone who read the original version of this article carried away a memory of a nice musky fragrance, an effect that Harvard University Press has wisely decided not to try to reproduce.

On October 15, 1764, Edward Gibbon conceived the idea of writing the history of the decline and fall of the Roman Empire while he was listening to barefoot monks singing vespers in the ruins of the Roman Capitol. I wish I could say I worked in settings that glamorous. I got the idea for my best-known work while I was driving my red Camaro in Cambridge, Massachusetts, on the way to my office in the physics department at the Massachusetts Institute of Technology.

I was feeling strung out. I had taken a leave of absence from my regular professorship at Berkeley a year earlier so that my wife could study at Harvard Law School. We had just gone through the trauma of moving from one rented house in Cambridge to another, and I had taken over the responsibility of getting our daughter to nursery school, playgrounds, and all that. More to the point, I was also stuck in my work as a theoretical physicist.

Like other theorists, I work with just pencil and paper, trying to make simple explanations of complicated phenomena. We leave it to the experimental physicists to decide whether our theories actually describe the real world. It was this opportunity to explain something about nature by noodling around with mathematical ideas that drew me into theoretical physics in the first place. For the previous two years, I had made progress in understanding what physicists call the strong interactions—the forces that hold particles together inside atomic nuclei. Some of my calculations had even been confirmed by experiments. But now these ideas seemed to be leading to nonsense. The new theories of the strong interactions I had been playing with that autumn implied that one of the particles of high energy nuclear physics should have no mass at all, but this particle was known to be actually quite heavy. Making predictions that are already known to be wrong is no way to get ahead in the physics game.

Often, when you're faced with a contradiction like this, it does no good to sit at your desk doing calculations—you just go round and round in circles. What does sometimes help is to let the problem cook on your brain's back burner while you sit on a park bench and watch your daughter play in a sandbox.

After this problem had been cooking in my mind for a few weeks, suddenly on my way to MIT (on October 2, 1967, as near as I can remember), I realized there was nothing wrong with the sort of theory on which I had been working. I had the right answer, but I had been working on the wrong problem. The mathematics I had been playing with had nothing to do with the strong interactions, but it gave a beautiful description of a different kind of force, known as the weak interaction. This is the force that is responsible, among other things, for the first step in the chain of nuclear reactions that produces the heat of the sun. There were inconsistencies in all previous theories of this force, but suddenly I saw how they could be solved. And I realized the massless particle in this theory that had given me so much trouble had nothing to do with the heavy particles that feel the strong interaction; it was the photon, the particle of which light is composed, that is responsible for electric and magnetic forces and that indeed has zero mass. I realized that what I had cooked up was an approach not just to understanding the weak interactions but to unifying the theories of the weak and electromagnetic forces into what has since come to be called the electroweak theory. This is just the sort of thing physicists love—to see several things that appear different as various aspects of one underlying phenomenon. Unifying the weak and electromagnetic forces might not have applications in medicine or technology, but if successful, it would be one more step in a centuries-old process of showing that nature is governed by simple, rational laws.

Somehow, I got safely to my office and started to work out the details of the theory. Where before I had been going around in circles, now everything was easy. Two weeks later, I mailed a short article on the electroweak theory to *Physical Review Letters,* a journal widely read by physicists.

The theory was proved to be consistent in 1971. Some new effects predicted by the theory were detected experimentally in 1973. By 1978, it was clear that measurements of these effects agreed precisely with the theory. And in 1979, I received the Nobel Prize in physics, along with Sheldon Glashow and Abdus Salam, who had done independent work on the electroweak theory. I have since learned that the paper I wrote in October 1967 has become the most cited article in the history of elementary particle physics.

I kept my red Camaro until it was totaled by one too many Massachusetts winters, but it never again took me so far.

17

The Non-Revolution
of Thomas Kuhn

In 1981 the mathematician Solomon Bochner founded an organization at Rice University called Scientia. His aim was to promote dialogue between scientists and engineers on the one hand, and humanists and social scientists on the other, an aim that has been pursued chiefly through an annual colloquium series. Scientia decided to devote this series for the 1997–98 academic year to the work of the celebrated historian and philosopher of science Thomas Kuhn. Some members had noticed the remarks about Kuhn's work that I had made in my *New York Review of Books* article about Sokal's hoax, reprinted in this collection, and asked me to give the annual Bochner Lecture as part of this series. After giving the talk, I edited the transcript into readable English and sent it to Robert Silvers, who agreed to publish it in *The New York Review of Books*.

This article prompted an unusually large number of comments. The one letter that was selected for publication in *The New York Review of Books* was answered in the short essay reprinted next. I received a number of other letters from philosophers and historians who generally agreed with what I had to say about Kuhn's theory of scientific revolutions, but pointed out that they or others had made similar criticisms earlier. I was glad to learn that I wasn't alone, and didn't really mind being scooped, since I don't set up to be a historian or philosopher. I also had the pleasure of seeing my remark about T-shirts and Maxwell's equations reprinted in *The Guardian* of London.

The most surprising comment came from the physicist Brian

Josephson, responding to a favorable editorial on my remarks about Kuhn that had appeared in the British periodical *Physics World*. Among his other objections, Josephson disagreed with my deemphasis of the degree to which scientists are hypnotized by their paradigms, and gave as an example of hypnosis by paradigm the dismissive things that I had said in my book *Dreams of a Final Theory* about research on telepathy. I replied that, in deciding not to study the evidence for telepathy, I was making the sort of judgment about how to spend one's working hours that everyone, including Josephson, must make. I went on to bet that Josephson had not carefully examined the evidence for astrology, and just to hedge that bet, I added that in any case he could not also have carefully examined the evidence for necromancy or the liquefaction of the blood of Saint Januarius.

In this essay I took the opportunity to answer some adverse comments by the philosopher Richard Rorty, responding to my remark about the reality of the laws of nature in my article "Sokal's Hoax." I found out about Rorty's comments by accident, since he had not sent me a copy of the article in which they were published. I have to admit that I didn't send him a copy of my answer either, assuming that he probably was a regular reader of *The New York Review of Books*. Perhaps we will continue this exchange by putting messages in bottles that we throw into the sea.

I first read Thomas Kuhn's famous book *The Structure of Scientific Revolutions*[1] a quarter-century ago, soon after the publication of the second edition. I had known Kuhn only slightly when we had been together on the faculty at Berkeley in the early 1960s, but I came to like and admire him later, when he came to MIT. His book I found exciting.

Evidently others felt the same. *Structure* has had a wider influence than any other book on the history of science. Soon after

1. Thomas S. Kuhn, *The Structure of Scientific Revolutions* (University of Chicago Press, 1962; 2nd ed., 1970), referred to in the text as *Structure*.

Kuhn's death in 1996, the sociologist Clifford Geertz remarked that Kuhn's book had "opened the door to the eruption of the sociology of knowledge" into the study of the sciences. Kuhn's ideas have been invoked again and again in the recent conflict over the relation of science and culture known as the science wars.

Structure describes the history of science as a cyclic process. There are periods of "normal science" that are characterized by what Kuhn sometimes called a "paradigm" and sometimes called a "common disciplinary matrix." Whatever you call it, it describes a consensus view: in a period of normal science, scientists tend to agree about what phenomena are relevant and what constitutes an explanation of these phenomena, about what problems are worth solving and what is a solution of a problem. Near the end of a period of normal science a crisis occurs—experiments give results that don't fit existing theories, or internal contradictions are discovered in these theories. There is alarm and confusion. Strange ideas fill the scientific literature. Eventually there is a revolution. Scientists become converted to a new way of looking at nature, resulting eventually in a new period of normal science. The "paradigm" has shifted.

To take an example given special attention in *Structure,* after the widespread acceptance of Newton's physical theories—the Newtonian paradigm—in the eighteenth century, there began a period of normal science in the study of motion and gravitation. Scientists used Newtonian theory to make increasingly accurate calculations of planetary orbits, leading to spectacular successes like the prediction in 1846 of the existence and orbit of the planet Neptune before astronomers discovered it. By the end of the nineteenth century there was a crisis: a failure to understand the motion of light. This problem was solved through a paradigm shift, a revolutionary revision in the understanding of space and time carried out by Einstein in the decade between 1905 and 1915, going far beyond the crisis that had inspired it. Motion affects the flow of time; matter and energy can he converted into each other; and gravitation is a curvature in spacetime. Einstein's Theory of Relativity then became the new paradigm, and the study of motion and gravitation entered upon a new period of normal science.

Though one can question the extent to which Kuhn's cyclic theory of scientific revolution fits what we know of the history of science, in itself this theory would not be very disturbing, nor would it have made Kuhn's book famous. For many people, it is Kuhn's reinvention of the word "paradigm" that has been either most useful or most objectionable. Of course, in ordinary English the word "paradigm" means some accomplishment that serves as a model for future work. This is the way that Kuhn had used this word in his earlier book[2] on the scientific revolution associated with Copernicus, and one way that he continued occasionally to use it.

The first critic who took issue with Kuhn's new use of the word "paradigm" in *Structure* was Harvard President James Bryant Conant. Kuhn had begun his career as a historian as Conant's assistant in teaching an undergraduate course at Harvard, when Conant asked Kuhn to prepare case studies on the history of mechanics. After seeing a draft of *Structure,* Conant complained to Kuhn that "paradigm" was "a word you seem to have fallen in love with!" and "a magical verbal word to explain everything!" A few years later Margaret Masterman pointed out that Kuhn had used the word "paradigm" in over twenty different ways.

But the quarrel over the word "paradigm" seems to me unimportant. Kuhn was right that there is more to a scientific consensus than just a set of explicit theories. We need a word for the complex of attitudes and traditions that go along with our theories in a period of normal science, and "paradigm" will do as well as any other.

What does bother me on rereading *Structure* and some of Kuhn's later writings is his radically skeptical conclusions about what is accomplished in the work of science.[3] And it is just these

2. Thomas S. Kuhn, *The Copernican Revolution* (Harvard University Press, 1957).

3. Kuhn was first trained as a physicist, and despite the presence of the wide-ranging word "scientific" in its title, *The Structure of Scientific Revolutions* is almost entirely concerned with physics and allied physical sciences like astron-

conclusions that have made Kuhn a hero to the philosophers, historians, sociologists, and cultural critics who question the objective character of scientific knowledge, and who prefer to describe scientific theories as social constructions, not so different in this respect from democracy or baseball.

Kuhn made the shift from one paradigm to another seem more like a religious conversion than an exercise of reason. He argued that our theories change so much in a paradigm shift that it is nearly impossible for scientists after a scientific revolution to see things as they had been seen under the previous paradigm. Kuhn compared the shift from one paradigm to another to a gestalt flip, like the optical illusion created by pictures in which what had seemed to be white rabbits against a black background suddenly appears as black goats against a white background. But for Kuhn the shift is more profound; he added that "the scientist does not preserve the gestalt subject's freedom to switch back and forth between ways of seeing."

Kuhn argued further that in scientific revolutions it is not only our scientific theories that change but the very standards by which scientific theories are judged, so that the paradigms that govern successive periods of normal science are *incommensurable*. He went on to reason that since a paradigm shift means complete abandonment of an earlier paradigm, and there is no common standard to judge scientific theories developed under different paradigms, there can be no sense in which theories developed after a scientific revolution can be said to add cumulatively to what was known before the revolution. Only within the context of a paradigm can we speak of one theory being true or false. Kuhn in *Structure* concluded, tentatively, "We may, to be more precise, have to relinquish the notion explicit or implicit that changes of paradigm carry scientists and those who learn from them closer and closer to the truth." More recently, in his Rothschild Lecture at Harvard in 1992, Kuhn remarked that it is hard to imagine

---

omy and chemistry. It is Kuhn's view of their history that I will be criticizing. I don't know enough about the history of the biological or behavioral sciences to judge whether anything I will say here also applies to them.

what can be meant by the phrase that a scientific theory takes us "closer to the truth."

Kuhn did not deny that there is progress in science, but he denied that it is progress *toward* anything. He often used the metaphor of biological evolution: scientific progress for him was like evolution as described by Darwin, a process driven from behind, rather than pulled toward some fixed goal to which it grows ever closer. For him, the natural selection of scientific theories is driven by problem solving. When, during a period of normal science, it turns out that some problems can't be solved by using existing theories, then new ideas proliferate, and the ideas that survive are those that do best at solving these problems. But according to Kuhn, just as there was nothing inevitable about mammals appearing in the Cretaceous period and out-surviving the dinosaurs when a comet hit the earth, so also there's nothing built into nature that made it inevitable that our science would evolve in the direction of Maxwell's equations or General Relativity. Kuhn recognizes that Maxwell's and Einstein's theories are better than those that preceded them, in the same way that mammals turned out to be better than dinosaurs at surviving the effects of comet impacts, but when new problems arise they will be replaced by new theories that are better at solving *those* problems, and so on, with no overall improvement.

All this is wormwood to scientists like myself, who think the task of science is to bring us closer and closer to objective truth. But Kuhn's conclusions are delicious to those who take a more skeptical view of the pretensions of science. If scientific theories can be judged only within the context of a particular paradigm, then in this respect the scientific theories of any one paradigm are not privileged over other ways of looking at the world, such as shamanism or astrology or creationism. If the transition from one paradigm to another cannot be judged by any external standard, then perhaps it is culture rather than nature that dictates the content of scientific theories.

Kuhn himself was not always happy with those who invoked

his work. In 1965 he complained that for the philosopher Paul Feyerabend to describe his arguments as a defense of irrationality in science seemed to him to be "not only absurd but vaguely obscene." In a 1991 interview with John Horgan, Kuhn sadly recalled a student in the 1960s complimenting him, "Oh, thank you, Mr. Kuhn, for telling us about paradigms. Now that we know about them, we can get rid of them." Kuhn was also uncomfortable with the so-called Strong Program in the sociology of science, which is "strong" in its uncompromisingly skeptical aim to show how political and social power and interests dominate the success or failure of scientific theories. This program is particularly associated with a group of philosophers and sociologists of science that at one time worked at the University of Edinburgh. About this, Kuhn remarked in 1991, "I am among those who have found the claims of the strong program absurd, an example of deconstruction gone mad."

But even when we put aside the excesses of Kuhn's admirers, the radical part of Kuhn's theory of scientific revolutions is radical enough. And I think it is quite wrong.

First, it is not true that scientists are unable to "switch back and forth between ways of seeing," and that after a scientific revolution they become incapable of understanding the science that went before it. As I said, one of the paradigm shifts to which Kuhn gives much attention in *Structure* is the replacement at the beginning of this century of Newtonian mechanics by the relativistic mechanics of Einstein. But in fact in educating new physicists the first thing that we teach them is still good old Newtonian mechanics, and they never forget how to think in Newtonian terms, even after they learn about Einstein's Theory of Relativity. Kuhn himself as an instructor at Harvard must have taught Newtonian mechanics to undergraduates.

In defending his position, Kuhn argued that the words we use and the symbols in our equations mean different things before and after a scientific revolution; for instance, physicists meant different things by mass before and after the advent of relativity. It is true that there was a good deal of uncertainty about the concept of mass *during* the Einsteinian revolution. For a while there was talk of "longitudinal" and "transverse" masses, which were supposed to depend on a particle's speed and to resist accelerations along the direction of motion and perpendicular to it. But this has all been resolved. No one today talks of longitudinal or transverse mass, and in fact the term "mass" today is most frequently understood as "rest mass," an intrinsic property of a body that is not changed by motion, which is much the way that mass was understood before Einstein. Meanings can change, but generally they do so in the direction of an increased richness and precision of definition, so that we do not lose the ability to understand the theories of past periods of normal science.

Perhaps Kuhn came to think that scientists in one period of normal science generally do not understand the science of earlier periods because of his experience in teaching and writing about the history of science. He probably had to contend with the ahistorical notions of scientists and students, who have not read original sources, and who believe that we can understand the work of the scientists in a revolutionary period by supposing that scientists of the past thought about their theories in the way that we describe these theories in our science textbooks. Kuhn's 1978 book[4] on the birth of quantum theory convinced me that I made just this mistake in trying to understand what Max Planck was doing when he introduced the idea of the quantum.

It is also true that scientists who come of age in a period of normal science find it extraordinarily difficult to understand the work of the scientists in previous scientific *revolutions,* so that in this respect we are often almost incapable of reliving the "gestalt flip" produced by the revolution. For instance, it is not easy for a physicist today to read Newton's *Principia,* even in a modern translation from Newton's Latin. The great astrophysicist Subrahmanyan

---

4. Thomas S. Kuhn, *Black-Body Theory and the Quantum Discontinuity,* 1894–1912 (Oxford University Press, 1978).

Chandrasekhar spent years translating the *Principia*'s reasoning into a form that a modern physicist could understand. But those who participate in a scientific revolution are in a sense living in two worlds: the earlier period of normal science, which is breaking down, and the new period of normal science, which they do not yet fully comprehend. It is much less difficult for scientists in one period of normal science to understand theories that have reached their mature form in an earlier period of normal science.

I was careful earlier to talk about Newtonian mechanics, not Newton's mechanics. In an important sense, especially in his geometric style, Newton is pre-Newtonian. Recall the aphorism of John Maynard Keynes, that Newton was not the first modern scientist but rather the last magician. Newtonianism reached its mature form in the early nineteenth century through the work of Pierre Simon de Laplace, Joseph Louis Lagrange, and others, and it is this mature Newtonianism—which still predates Special Relativity by a century—that we teach our students today. They have no trouble in understanding it, and they continue to understand it and use it where appropriate after they learn about Einstein's Theory of Relativity.

Much the same could be said about our understanding of the electrodynamics of James Clerk Maxwell. Maxwell's 1873 *Treatise on Electricity and Magnetism* is difficult for a modern physicist to read, because it is based on the idea that electric and magnetic fields represent tensions in a physical medium, the ether, in which we no longer believe. In this respect, Maxwell is pre-Maxwellian. (Oliver Heaviside, who helped to refine Maxwell's theory, said of Maxwell that he was only half a Maxwellian.) Maxwellianism—the theory of electricity, magnetism, and light that is based on Maxwell's work—reached its mature form (which does not require reference to an ether) by 1900, and it is this *mature* Maxwellianism that we teach our students. Later they take courses on quantum mechanics in which they learn that light is composed of particles called photons, and that Maxwell's equations are only approximate; but this does not prevent them from continuing to understand and use Maxwellian electrodynamics where appropriate.

In judging the nature of scientific progress, we have to look at mature scientific theories, not theories at the moments when they are coming into being. If it made sense to ask whether the Norman Conquest turned out to be a good thing, we might try to answer the question by comparing Anglo-Saxon and Norman societies in their mature forms—say, in the reigns of Edward the Confessor and Henry I. We would not try to answer it by studying what happened at the Battle of Hastings.

Nor do scientific revolutions necessarily change the way that we assess our theories, making different paradigms incommensurable. Over the past forty years I have been involved in revolutionary changes in the way that physicists understand the elementary particles that are the basic constituents of matter. The greater revolutions of this century, quantum mechanics and relativity, were before my time, but they are the basis of the physics research of my generation. Nowhere have I seen any signs of Kuhn's incommensurability between different paradigms. Our ideas have changed, but we have continued to assess our theories in pretty much the same way: a theory is taken as a success if it is based on simple general principles and does a good job of accounting for experimental data in a natural way. I am not saying that we have a book of rules that tells us how to assess theories, or that we have a clear idea what is meant by "simple general principles" or "natural." I am only saying that whatever we mean, there have been no sudden changes in the way we assess theories that would make it impossible to compare the truth of theories before and after a revolution.

For instance, at the beginning of this century physicists were confronted with the problem of understanding the spectra of atoms, the huge number of bright and dark lines that appear in the light from hot gases, like those on the surface of the sun, when the light is spread out by a spectroscope into its different colors. When Niels Bohr showed in 1913 how to use quantum theory to explain the spectrum of hydrogen, it became clear to physicists generally that quantum theory was very promising, and when it turned out after 1925 that quantum mechanics could be used to explain the spectrum of any atom, quantum mechanics became the hot subject that young physicists had to learn. In the same way, physicists today are confronted with a dozen or so measured masses for the electron and similar particles and for quarks of various types, and the measured numerical values of these different masses have so far resisted theoretical explanation. Any new theory that succeeds in explaining these masses will instantly be recognized as an important step forward. The subject matter has changed, but not our aims.

This is not to say that there have been no changes at all in the way we assess our theories. For instance, it is now considered to be much more acceptable to base a physical theory on some principle of "invariance" (a principle that says that the laws of nature appear the same from certain different points of view) than it was at the beginning of the century, when Einstein started to worry about the invariance of the laws of nature under changes in the motion of an observer. But these changes have been evolutionary, not revolutionary. Nature seems to act on us as a teaching machine. When a scientist reaches a new understanding of nature, he or she experiences an intense pleasure. These experiences over long periods have taught us how to judge what sort of scientific theory will provide the pleasure of understanding nature.

Even more radical than Kuhn's notion of the incommensurability of different paradigms is his conclusion that in the revolutionary shifts from one paradigm to another we do not move closer to the truth. To defend this conclusion, he argued that all past beliefs about nature have turned out to be false, and that there is no reason to suppose that we are doing better now. Of course, Kuhn knew very well that physicists today go on using the Newtonian theory of gravitation and motion and the Maxwellian theory of electricity and magnetism as good approximations that can be deduced from more accurate theories—we certainly don't regard Newtonian and Maxwellian theories as simply false, in the way that Aristotle's theory of motion or the theory that fire is an element ("phlogiston") are false. Kuhn himself in his earlier book on the Copernican revolution told how parts of scientific theories survive in the more successful theories that supplant them, and seemed to have no trouble with the idea. Confronting this contradiction, Kuhn in *Structure* gave what for him was a remarkably weak defense, that Newtonian mechanics and Maxwellian electrodynamics as we use them today are not the same theories as they were before the advent of relativity and quantum mechanics, because then they were not known to be approximate and now we know that they are. It is like saying that the steak you eat is not the one that you bought, because now you know it is stringy and before you didn't.

It is important to keep straight what does and what does not change in scientific revolutions, a distinction that is not made in *Structure*.[5] There is a "hard" part of modern physical theories ("hard" meaning not difficult, but durable, like bones in paleontology or potsherds in archeology) that usually consists of the equations themselves, together with some understandings about what the symbols mean operationally and about the sorts of phenomena to which they apply. Then there is a "soft" part; it is the vision of reality that we use to explain to ourselves why the equations work. The soft part does change; we no longer believe in Maxwell's ether, and we know that there is more to nature than Newton's particles and forces.

The changes in the soft part of scientific theories also produce changes in our understanding of the conditions under which the hard part is a good approximation. But after our theories reach their mature forms, their hard parts represent permanent accomplishments. If you have bought one of those T-shirts with Maxwell's equations on the front, you may have to worry about its going out of style, but not about its becoming false. We will go

---

5. I am grateful to Professor Christopher Hitchcock for a comment after my talk at Rice that led me to include the following remark in this essay.

on teaching Maxwellian electrodynamics as long as there are scientists. I can't see any sense in which the increase in scope and accuracy of the hard parts of our theories is *not* a cumulative approach to truth.[6]

Some of what Kuhn said about paradigm shifts does apply to the soft parts of our theories, but even here I think that Kuhn overestimated the degree to which scientists during a period of normal science are captives of their paradigms. There are many examples of scientists who remained skeptical about the soft parts of their own theories. It seems to me that Newton's famous slogan *Hypotheses non fingo* (I do not make hypotheses) must have meant at least in part that his commitment was not to the reality of gravitational forces acting at a distance, but only to the validity of the predictions derived from his equations.

However that may be, I can testify that although our present theory of elementary particles, the Standard Model, has been tremendously successful in accounting for the measured properties of the known particles, physicists today are not firmly committed to the view of nature on which it is based. The Standard Model is a field theory, which means that it takes the basic constituents of nature to be fields—conditions of space, considered apart from any matter that may be in it, like the magnetic field that pulls bits of iron toward the poles of a bar magnet—rather than particles. In the past two decades it has been realized that any theory based on quantum mechanics and relativity will look like a field theory

---

6. Another complication: As Professor Bruce Hunt pointed out to me in conversation, it can happen that two competing theories with apparently different hard parts can both make the same successful predictions. For instance, in the mid-nineteenth century it was common for British physicists to describe electromagnetic phenomena using equations that involved electric and magnetic fields, following the lead of Faraday, while the equations of continental physicists referred directly to forces acting at a distance. Usually what happens in such cases is that the two sets of equations are discovered to be mathematically equivalent, although one or the other may turn out to have a wider generalization in a more comprehensive theory, as turned out to be the case for electric and magnetic fields after the advent of Maxwell's theory.

when experiments are done at sufficiently low energies. The Standard Model is today widely regarded as an "effective field theory," a low energy approximation to some unknown fundamental theory that may not involve fields at all.

Even though the Standard Model provides the paradigm for the present normal-science period in fundamental physics, it has several ad hoc features, including at least eighteen numerical constants, such as the mass and charge of the electron, that have to be arbitrarily adjusted to make the theory fit experiments. Also, the Standard Model does not incorporate gravitation. Theorists know that they need to find a more satisfying new theory, to which the Standard Model would be only a good approximation, and experimentalists are working very hard to find some new data that would disagree with some prediction of the Standard Model. The recent announcement from an underground experiment in Japan, that the particles called neutrinos have masses that would be forbidden in the original version of the Standard Model, provides a good example. This experiment is only the latest step in a search over many years for such masses, a search that has been guided in part by arguments that, whatever more satisfying theory turns out to be the next step beyond the Standard Model, this theory is likely to entail the existence of small neutrino masses.

Kuhn overstated the degree to which we are hypnotized by our paradigms, and in particular he exaggerated the extent to which the discovery of anomalies during a period of normal science is inadvertent. He was quite wrong in saying that it is no part of the work of normal science to find new sorts of phenomena.

To adopt Kuhn's view of scientific progress would leave us with a mystery: Why does anyone bother? If one scientific theory is only better than another in its ability to solve the problems that happen to be on our minds today, then why not save ourselves a lot of trouble by putting these problems out of our minds? We don't study elementary particles because they are intrinsically interesting, like people. They are not—if you have seen one electron, you've seen them all. What drives us onward in the work of science is precisely the sense that there are truths out there to be discovered, truths that once discovered will form a permanent part of human knowledge.

It was not Kuhn's description of scientific revolutions that impressed me so much when I first read *Structure* in 1972, but rather his treatment of normal science. Kuhn showed that a period of normal science is not a time of stagnation, but an essential phase of scientific progress. This had become important to me personally in the early 1970s because of recent developments in both cosmology and elementary particle physics.

Until the late 1960s cosmology had been in a state of terrible confusion. I remember when most astronomers and astrophysicists were partisans of some preferred cosmology, and considered anyone else's cosmology mere dogma. Once at a dinner party in New York around 1970 I was sitting with the distinguished Swedish physicist Hannes Alfven, and took the opportunity to ask whether or not certain physical effects on which he was an expert would have occurred in the early universe. He asked me, "Is your question posed within the context of the Big Bang theory?" and when I said yes, it was, he said that then he didn't want to talk about it. The fractured state of cosmological discourse began to heal after the discovery in 1965 of the cosmic microwave background radiation, radiation that is left over from the time when the universe was about a million years old. This discovery forced everyone (or at least almost everyone) to think seriously about the early universe.

At last measurements were being made that could confirm or refute our cosmological speculations, and very soon, in less than a decade, the Big Bang theory was developed in its modern form and became widely accepted. In a treatise on gravitation and cosmology that I finished in 1971 I introduced the phrase "Standard Model" for the modern Big Bang cosmology, to emphasize that I regarded it not as a dogma to which everyone had to swear allegiance, but as a common ground on which all physicists and astronomers could meet to discuss cosmological calculations and observations. There remained respected physicists and astronomers, like Alfven and Fred Hoyle, who did not like the direction of the growing consensus. Some of them attacked the very idea of consensus, holding out instead a sort of "Shining Path" ideal of science as a continual revolution, in which all should pursue their own ideas and go off in their own directions. But there is much more danger in a breakdown of communication among scientists than in a premature consensus that happens to be in error. It is only when scientists share a consensus that they can focus on the experiments and the calculations that can tell them whether their theories are right or wrong, and, if wrong, can show the way to a new consensus. It was to good effect that Kuhn quoted Francis Bacon's dictum "Truth emerges more readily from error than from confusion."

Elementary particle physics also was entering into a new period of normal science the early 1970s. It had earlier been in a state of confusion, not because of a lack of data, of which there was more than enough, but because of the lack of a convincing body of theory that could explain this data. By the early 1970s theoretical developments and some important new experiments led to a consensus among elementary particle physicists, embodied in what is now also called a Standard Model. Yet for a while some physicists remained skeptical because they felt there hadn't been enough experiments done yet to prove the correctness of the Standard Model, or that the experimental data could be interpreted in other ways. When I argued that any other way of interpreting the data was ugly and artificial, some physicists answered that science has nothing to do with aesthetic judgments, a response that would have amused Kuhn. As he said, "The act of judgment that leads scientists to reject the previously accepted theory is always based upon more than a comparison of that theory with the world." Any set of data can be fit by many different theories. In deciding among these theories we have to judge which ones have the kind of elegance and consistency and universality that make them worth taking seriously. Kuhn was by no means the first person to make this point—he was preceded by, among others, Pierre Duhem—but Kuhn made it very convincingly.

By now the arguments about the Standard Model are pretty well over, and it is almost universally agreed to give a correct account of observed phenomena. We are living in a new period of normal science, in which the implications of the Standard Model are being calculated by theorists and tested by experimentalists. As Kuhn recognized, it is precisely this sort of work during periods of normal science that can lead to the discovery of anomalies that will make it necessary to take the next step beyond our present paradigm.

But Kuhn's view of normal science, though it remains helpful and insightful, is not what made his reputation. The famous part of his work is his description of scientific revolutions and his view of scientific progress. And it is here that his work is so seriously misleading.

What went wrong? What in Kuhn's life led him to his radical skepticism, to his strange view of the progress of science? Certainly not ignorance—he evidently understood many episodes in the history of physical science as well as anyone ever has. I picked up a clue to Kuhn's thinking the last time I saw him, at a ceremony in Padua in 1992 celebrating the four hundredth anniversary of the first lecture Galileo delivered in the University of Padua. Kuhn told how in 1947 as a young physics instructor at Harvard, studying Aristotle's work in physics, he had been wondering:

> How could [Aristotle's] characteristic talent have deserted him so systematically when he turned to the study of motion and mechanics? Equally, if his talents had deserted him, why had his writings in physics been taken so seriously for so many centuries after his death? . . . Suddenly the fragments in my head sorted themselves out in a new way, and fell into place altogether. My jaw dropped with surprise, for all at once Aristotle seemed a very good physicist indeed, but of a sort I'd never dreamed possible.

I asked Kuhn what he had suddenly understood about Aristotle. He didn't answer my question, but wrote to me to tell me again how important this experience was to him:

> What was altered by my own first reading of [Aristotle's writings on physics] was my understanding, not my evaluation, of what they achieved. And what made that change an epiphany was the transformation it immediately effected in my understanding (again, not my evaluation) of the nature of scientific achievement, most immediately the achievements of Galileo and Newton.

Later, I read Kuhn's explanation in a 1977 article that, without becoming an Aristotelian physicist, he had for a moment learned to think like one, to think of motion as a change in the quality of an object that is like many other changes in quality rather than a state that can be studied in isolation. This apparently showed Kuhn how with sufficient effort it is possible to adopt the point of view of any scientist one studies. I suspect that because this moment in his life was so important to Kuhn, he took his idea of a paradigm shift from the shift from Aristotelian to Newtonian physics—the shift (which actually took many centuries) from Aristotle's attempt to give systematic qualitative descriptions of everything in nature to Newton's quantitative explanations of carefully selected phenomena, such as the motion of the planets around the sun.

Now, that really *was* a paradigm shift. For Kuhn it seems to have been the paradigm of paradigm shifts, which set a pattern into which he tried to shoehorn every other scientific revolution. It really does fit Kuhn's description of paradigm shifts: it is extraordinarily difficult for a modern scientist to get into the frame of mind of Aristotelian physics, and Kuhn's statement that all previous views of reality have proved false, though not true of Newtonian mechanics or Maxwellian electrodynamics, certainly does apply to Aristotelian physics.

Revolutions in science seem to fit Kuhn's description only to the extent that they mark a shift in understanding some aspect of

nature from pre-science to modern science. The birth of Newtonian physics was a mega-paradigm shift, but nothing that has happened in our understanding of motion since then—not the transition from Newtonian to Einsteinian mechanics, or from classical
to quantum physics—fits Kuhn's description of a paradigm shift.

During the last few decades of his life Kuhn worked as a philosopher, worrying about the meaning of truth and reality, problems
on which he had touched briefly decades earlier in *Structure*. After
Kuhn's death Richard Rorty said that Kuhn was "the most influential philosopher to write in English since World War II." Kuhn's
conclusions about philosophy show the same corrosive skepticism
as his writings on history. In his Rothschild Lecture at Harvard in
1992, he remarked, "I am not suggesting, let me emphasize, that
there is a reality which science fails to get at. My point is rather
that no sense can be made of the notion of reality as it has ordinarily functioned in the philosophy of science."

It seems to me that pretty good sense had been made of the notion of reality over a century ago by the pragmatic philosopher
Charles Sanders Peirce, but I am not equipped by taste or education to judge conflicts among philosophers. Fortunately we need
not allow philosophers to dictate how philosophical arguments
are to be applied in the history of science, or in scientific research
itself, any more than we would allow scientists to decide by themselves how scientific discoveries are to be used in technology or
medicine.

I remarked in a recent article in *The New York Review of Books*
that for me as a physicist the laws of nature are real in the same
sense (whatever that is) as the rocks on the ground.[7] A few months
after the publication of my article I was attacked for this remark
by Richard Rorty.[8] He accused me of thinking that as a physicist I
can easily clear up questions about reality and truth that have engaged philosophers for millennia. But that is not my position. I

7. "Sokal's Hoax," Essay 9 above.
8. Richard Rorty, "Thomas Kuhn, Rocks, and the Laws of Physics," *Common Knowledge*, 6, no. 1 (1997), 6.

know that it is terribly hard to say precisely what we mean when we use words like "real" and "true." That is why, when I said that the laws of nature and the rocks on the ground are real in the same sense, I added in parentheses "whatever that is." I respect the efforts of philosophers to clarify these concepts, but I'm sure that even Kuhn and Rorty have used words like "truth" and "reality" in everyday life, and had no trouble with them. I don't see any reason why we cannot also use them in talking about the history of science. Certainly philosophers can do us a great service in their attempts to clarify what we mean by truth and reality. But for Kuhn to say that as a philosopher he has trouble understanding what is meant by truth or reality proves nothing beyond the fact that he has trouble understanding what is meant by truth or reality.

Finally, I would like to describe my own idea of scientific progress. As I said, Kuhn uses the metaphor of Darwinian evolution: undirected improvement, but not improvement toward anything. Kuhn's metaphor is not bad, if we make one change in it: the progress of physical science looks like evolution running backward. Just as humans and other mammal species can trace their origins back to some kind of furry creature hiding from the dinosaurs in the Cretaceous period, and that furry creature and the dinosaurs and all life on earth presumably can be traced back to what Pooh-Bah in *The Mikado* called "a protoplasmal primordial atomic globule," so correspondingly we have seen the science of optics and the science of electricity and magnetism merge together in Maxwell's time into what we now call electrodynamics, and in recent years we have seen electrodynamics and the theories of other forces in nature merge into the modern Standard Model of elementary particles. We hope that in the next great step forward in physics we shall see the theory of gravitation and all of the different branches of elementary particle physics flow together into a single unified theory. This is what we are working for and what we spend the taxpayers' money for. And when we have discovered this theory, it will be part of a true description of reality.

18

T. S. Kuhn's Non-Revolution:
An Exchange

The New York Review of Books decided to publish one of the letters written in response to my article about Thomas Kuhn, a letter by Professor Alex Levine of the Lehigh University Department of Philosophy. In it Professor Levine claimed that my disagreements with Kuhn are based on common misconceptions about Kuhn's work that I share with many historians and philosophers, as well as fellow scientists. It was good to learn that I was not alone in my views, but I did not think that I had misunderstood Kuhn.

Professor Levine emphasized that Kuhn rejected the charge that his reading of history made science an irrational enterprise; he repeated Kuhn's explanation that Maxwell's equations mean something different now from what they meant in the past because now we know that they are only approximate; he argued that Aristotle's astronomy has survived in the same sense as Newton's laws and Maxwell's equations, because we still use it for celestial navigation; and he claimed that Kuhn's view of the history of science was more scientific than mine, because I appear to assume "that science must inevitably succeed" in coming closer and closer to objective truth, while Kuhn had treated this as a matter to be settled empirically. Here is my answer, as published along with Professor Levine's letter.

Have I misunderstood Kuhn? Of course Kuhn never claimed that science is an irrational enterprise. Nor have I said that he did. But I am glad that Professor Levine has given me another chance to comment on the topic of rationality, for it seems natural to me that Feyerabend and others should have seen Kuhn's arguments as a defense of irrationality in science. If I agreed with Kuhn's judgment about the progress of science, that there is no sense in which science offers a cumulative approach to some sort of truth, then the whole enterprise would seem rather irrational to *me*, even if not to Kuhn.

Regarding the falsity of past theories, I can't detect any difference between the way that Professor Levine and I describe Kuhn's views. The difference between Professor Levine and myself is not that I understand Kuhn differently, but that I disagree with Kuhn.

For one thing, I don't see the difficulty in describing an approximate theory as "true" or "false." As Professor Levine has guessed, I don't regard approximations as mere useful fictions. One of the problems that I have just assigned to the students in my class in quantum mechanics at the University of Texas calls for an approximate calculation of the shift in the energy of the hydrogen atom when it is placed in a weak electric field, neglecting various small effects in the atom. If they want full marks, they had better get an answer that is true within these approximations. If they don't, then I won't look kindly on any excuse that an approximation is just a useful fiction.

But approximate theories are not merely approximately true. They can make a statement that, though it refers to an approximation, is nevertheless precisely true. For instance, although Maxwell's equations give only an approximate account of electric and magnetic fields, it is precisely true that the error introduced by using Maxwell's equations to calculate these fields can be made as small as one likes by considering fields that are sufficiently weak and slowly varying. This is part of the reason that Maxwell's equations are a permanent part of physical science.

Then there is the quite different question, to what extent the scientists of the past suspected that their theories were only approximate. Doubtless some didn't, but many others did. From the eighteenth century on there were astronomers who speculated about possible departures from the Newtonian theory of gravitation. Take a look at the article on the planet Mercury in the eleventh edition of the *Encyclopedia Britannica*, written by the noted astronomer Simon Newcomb before the advent of Einstein's General Relativity. You will find Newcomb speculating that Newton's inverse square law of gravitation may be only approximate. Likewise, even in the golden age of Maxwellian electrodynamics at the end of the nineteenth century, physicists like Joseph Larmor and Oliver Heaviside (and perhaps Maxwell himself) expected that there should be small corrections to Maxwell's equations. Physicists today recognize that their most fundamental theories, such as General Relativity and the Standard Model of elementary particles, are only approximations to a more fundamental future theory.

Incidentally, I don't agree that today Aristotle's scientific theories are still employed as useful fictions, for his theories had no quantitative content whatever. Ptolemaic astronomy is a different story, but I didn't mention it in my article.

In his final paragraphs Professor Levine misquotes my criticism of Kuhn's view of scientific progress. I never argued that progress toward objective truth is *inevitable.* My difference with Kuhn is that I think that such progress has in fact happened again and again in the history of modern science. Our future progress toward a fundamental physical theory may be stopped if it turns out that humans are not smart enough to conceive of such a theory (which I doubt), or if we find an infinite regress of more and more fundamental theories (also unlikely), or if society stops providing the resources for continued experiments (all too possible). But although I can't claim to know that we will be able to continue our progress toward a simple, objectively true theory underlying all natural phenomena, I do think that we must act under this assumption, for if we do not then our progress will surely not continue.

19

The Great Reduction:
Physics in the Twentieth Century

By now the excitement of passing through the start of a new century—even one whose years begin with a 2 rather than a 1—has pretty well worn off. Still, the event offered publishers a rare opportunity to bring out books and articles looking back over the past century, or looking ahead to the next. Among these publishers was Oxford University Press. Michael Howard, the Regius Professor of History at Oxford, and W. Roger Louis, the Kerr Professor of English History at my own University of Texas at Austin and Fellow of Oxford's St. Antony's College, agreed to edit *The Oxford History of the Twentieth Century,* and assembled a group of historians to cover different aspects of the century's history. I am not a historian, but I like reading history and I try to bring a historical perspective even into my technical writing, so when I was asked to contribute an article on the history of science in the twentieth century I thought I might give it a try.

Unfortunately, I didn't know enough about the history of biology or earth sciences to write a comprehensive history of twentieth-century science. I decided instead to concentrate on what interested me most—the remarkable success in the past century of physics in the reductive mode, of explaining more and more phenomena in terms of simpler and simpler principles. The title "The Great Reduction" is in part an in-your-face affirmation of the reductionist aims of the sort of physics I was writing about, and in part a knock-off of the title of a book by the sinologist Joseph Needham that I had admired: *The Grand Titration.* (Reduction like titration is a term of chemistry.)

In writing this article I faced a problem: the achievements in physics of the past half century were mostly the work of people still alive, many of them good friends. I might either offend physicists by not mentioning their names or exasperate the general reader by mentioning too many names. So I took the coward's way out: for any work done after 1950 I gave no names at all.

By the end of the nineteenth century scientists had achieved a fair understanding of the world of everyday experience. Sciences such as mechanics, thermodynamics, chemistry, optics, cell biology, and even electrodynamics had become part of the armamentarium of industry and agriculture. In return, industry had provided science with the apparatus, the vacuum pumps and spectroscopes and so on, that would be needed for further advances. Science had also been the beneficiary of the growing intellectual technology of mathematics. Differential and integral calculus, complex analysis, differential geometry, and group theory had become well enough developed by 1900 to satisfy the needs of science for decades to come.

One great thing was missing from science at the nineteenth century's end: a sense of how it all fitted together. In 1900 scientists generally supposed that physics, chemistry, and biology each operated under its own autonomous laws. A few scientists held on to Newton's dream of a grand synthesis of all the sciences, but without any clear idea of the terms on which this synthesis could be reached.

We catch echoes of this confusion in the complacent statements of late-nineteenth-century physicists asserting the near completion of physics. Those who made such remarks could not have thought that physics had completed its task of explaining the rules of chemistry, or that physics and chemistry had succeeded in explaining the mechanisms of heredity and embryology. Rather, they did not suppose that it was part of the task of physics to explain chemistry, or of physical science to explain life. The complacency of late-nineteenth-century physicists is a measure of the limitations of their ambitions for physics.

Matters are very different now, at the end of the twentieth century. We understand in principle and often in practice how the laws of chemistry arise from the laws of physics. Biology, of course, involves historical accidents in a way that physics and chemistry do not, but hardly anyone would seriously consider the possibility that there are autonomous universal laws of biology, independent of those of physical science.

This has been reduction in the grand sense—the explanation of a wide range of scientific principles in terms of simpler, more universal ones. It has also largely been reduction in the petty sense— the explanation of natural phenomena at one scale in terms of the principles governing matter at much smaller scales. Just as the rules of chemistry have been explained in terms of the dynamics of electrons in atoms, the mechanism of heredity that drives biological evolution is now understood in molecular terms.

This essay will deal not with all of twentieth-century science, nor even all of twentieth-century physics, but only with that part of physics research that has followed the reductionist tradition of seeking the deepest principles underlying all of nature.

If only for physicists, the twentieth century began in 1895, with Wilhelm Röntgen's accidental discovery of x-rays. Röntgen found that the electric current driven by a powerful electric battery through an evacuated glass "Crookes" tube (the first particle accelerator) produced mysterious rays on striking the glass wall of the tube, rays that could not be refracted by lenses or deflected by matter or by electric or magnetic fields. In itself, this discovery marked no great revolution. We know now that these x-rays are nothing but light waves of short wavelength, but Röntgen's discovery was a dramatic sign that, outside the range of natural phenomena that had been illuminated by nineteenth-century physics, there were plenty of mysteries waiting to be discovered.

Almost immediately, the discovery of x-rays spurred other, more important, discoveries. Early in 1896 Henri Becquerel, searching for new kinds of rays, found by accident that uranium emits rays that can expose photographic plates. Pierre and Marie Curie soon discovered that thorium, polonium, and radium produce similar effects, and gave the phenomenon its modern name, radioactivity. After a good deal of confusion Ernest Rutherford and Frederick Soddy identified radioactivity as being due to transmutation of elements, in which energy is released and carried off by particles of various sorts. Beyond its importance for the understanding of matter, the discovery of radioactivity soon suggested solutions to the fundamental problems of the source of the heat of the sun and the earth, and thereby resolved the discrepancy between the short age of the earth previously inferred from its present heat and the much longer age that had been deduced from the fossil record.

Only two years after the discovery of x-rays, both J. J. Thomson in Cambridge and Walter Kaufmann in Berlin measured the ratio of mass to electric charge of the currents flowing in a modified Crookes tube. This mass/charge ratio turned out to be thousands of times smaller than that for the ions that carry electric currents in electrolysis. On this basis, Thomson proposed that the current in a Crookes tube is carried by particles that he called electrons, thousands of times lighter than whole atoms, that are universally present in all matter.

The idea that matter consists of atoms had proved useful to chemists and physicists of the nineteenth century, but there was still no direct evidence of their existence. This was provided in the first decade of the twentieth century by a variety of measurements of the masses of atoms, generally giving consistent results. Most of these new measurements had been made possible by the discoveries of 1895–1897.

Until 1911 nobody had any idea how masses and electric charges are arranged within atoms. Then Rutherford used the results of experiments in his Manchester laboratory, on the scattering of particles from a radioactive source by gold atoms, to show that the mass of an atom is almost entirely concentrated in a tiny, positively charged nucleus. He pictured the electrons that are responsible for chemical, electrical, and optical phenomena as revolving around the nucleus like planets revolving around the sun. The age-old question of the nature of ordinary matter had thus been answered, with only the composition of the atomic nucleus still entirely mysterious.

In the great work of 1895–1911 described so far, experimenters played the leading role. Where physical theory was used to analyze the experimental results, it was the familiar classical theory of the nineteenth century. But the picture of the atom that emerged from this work raised disturbing questions that could not be answered within the classical framework.

According to the principles of electrodynamics, atomic electrons revolving in their orbits should emit light, losing energy and quickly spiraling into the nucleus. Yet every ordinary nonradioactive atom has a state of lowest energy that seems perfectly stable. Furthermore, although an atom that has absorbed energy from light or from collisions with other atoms will lose it again by emitting light, this emitted light does not have a continuous spectrum of wavelengths, as would be expected for electrons spiraling inward to a state of lowest energy; rather, it is emitted only at certain sharp wavelengths that are characteristic of the type of atom. Such atomic spectra were discovered early in the nineteenth century, and had become a tool for chemical analysis, including even the analysis of the elements in the sun and other stars, but no one could explain the spectra of wavelengths at which various atoms emit light.

The solution lay not in a new theory of electromagnetism or atomic structure, but in a radically new framework for all physical theory, known as quantum mechanics. Here are the highlights of this revolution:

1900 MAX PLANCK explains the spectrum of black-body radiation (the intensity of light emitted at various wavelengths by a heated opaque body) by assuming that the energy of radiation comes in multiples of a "quantum" of energy, inversely proportional to the light wavelength.

1905 ALBERT EINSTEIN interprets Planck's conjecture (over Planck's objections) as showing that light itself consists of individual particles, later called photons. Photons have no mass but have energy and momentum, both inversely proportional to the light wavelength. Experimental evidence for Einstein's photon is found in the photoelectric effect, and later in the scattering of x-rays by electrons.

1913 NIELS BOHR offers a tentative interpretation of atomic spectra. Atoms can exist only in certain states of definite energy. When an atom drops from a state of higher to one of lower energy, it emits light with an energy equal to the difference of the energies of the final and initial atomic states, and hence, according to Einstein, with a definite wavelength, inversely proportional to this energy difference. Bohr uses the Rutherford picture of atomic electrons revolving around a nucleus to give a successful though ad hoc prescription for calculating the energies of the states of simple atoms like hydrogen, which has just a single electron. But it is not clear how to apply this prescription to more complicated atoms and molecules, or why it should work for any atoms.

1923 LOUIS DE BROGLIE offers an explanation of Bohr's prescription. Electrons are associated with waves, with a wavelength inversely proportional to the electron momentum, just as for photons. The Bohr prescription for the energy of an atomic state is interpreted as the requirement that a whole number of these wavelengths should just fit around the electron orbit. But, again, this approach works only for the simplest atoms.

1925 WERNER HEISENBERG makes a fresh start at interpreting atomic spectra. The dynamics of atoms is expressed not in terms of the trajectories of electrons revolving around the nucleus, but in "tables" (matrices) of numerical quantities whose squares give the rates at which atoms make transitions from one atomic state to another. Soon after, in a mathematical tour de force, Heisenberg's former classmate Wolfgang Pauli uses Heisenberg's formalism to calculate the spectrum of hydrogen, recovering Bohr's 1913 results. But it still seems hopeless to use these methods in more complicated problems.

1925–26 ERWIN SCHRÖDINGER returns to de Broglie's electron waves, and proposes a differential equation that governs the amplitude of the wave (the so-called wave function) in general electric fields. He shows how Heisenberg's matrices and the algebraic conditions they satisfy can be derived from his wave equation. For the first time, there is a method that can be used to calculate the energies of states not just in hydrogen, but in atoms and molecules of arbitrary complexity. But the physical significance of the electron wave function remains obscure.

1926 MAX BORN explains that the waves of de Broglie and Schrödinger are "probability amplitudes"; the probability that an electron will be found in a small volume around a given point equals the square of the wave function at that point, times the volume.

After 1926 quantum mechanics was rapidly and successfully applied to the calculation of the properties not only of atoms in isolation, but also of atoms joined in chemical molecules and even in macroscopic solids. It became clear that the rules of chemistry are not autonomous scientific principles, but mathematical consequences of quantum mechanics as applied to electrons and atomic nuclei. This was arguably the single greatest triumph of science in the reductionist mode. Paul Dirac expressed the exuberance of the times when, in 1929, he proclaimed that "the underlying physical laws necessary for the mathematical theory of a larger part of physics and the whole of chemistry are thus completely known, and the difficulty is only that the application of these laws leads to equations much too complicated to be soluble."

A price had to be paid for this success. There is an irreducible strangeness to quantum mechanics. A particle does not have a definite position or momentum until one or the other is measured; what it does have is a wave function. The wave function evolves deterministically—knowing the wave function at one moment, you can use the Schrödinger equation to calculate the wave function at any later time. But, knowing the wave function, you cannot predict the precise result of a measurement of position or momentum—only the probabilities of the various possible results. How do these probabilities arise, if observers and their measuring apparatus are themselves described by a complicated but deterministic Schrödinger equation?

The strangeness of quantum mechanics raised a barrier to communication not only between physicists and nonphysicists, but also between the young physicists of the 1920s and their older colleagues. Einstein and Planck never accepted the consensus version of quantum mechanics, and its meaning continued to be debated throughout the century. Yet quantum mechanics proved remarkably resilient. The triumphs of atomic, condensed matter, nuclear, and elementary particle physics from the 1920s to the present have been based on the quantum mechanical framework that was already in place by 1926.

The other great revolution of twentieth-century physics was the development of the Theory of Relativity. Although not as radical a break with the past as quantum mechanics, relativity was in one way more remarkable. Quantum mechanics was developed over a quarter-century by successive contributions of many physicists, while relativity was almost entirely due to the work of a single physicist, Albert Einstein, from 1905 to 1915.

From an early age, Einstein had worried about the effect of an observer's motion on observations of light. By the end of the nineteenth century this problem had become acute, with experimenters' efforts continually failing to detect changes in the speed of light caused by the earth's motion around the sun. Several theorists addressed this problem dynamically, through hypotheses about the effect of motion through the ether (the medium that was believed to carry light waves) on the measuring apparatus used to measure the speed of light. Then in 1905 Einstein put the ether aside, and asserted as a fundamental law of physics the Special Principle of Relativity—that no uniform motion of an observer could have any effect on the speed of light or anything else.

Relativity was not new. A principle of relativity had been asserted by Galileo, and was an important ingredient of Newton's mechanics; all unaccelerated observers who study the motions of masses moving under the influence of their mutual gravitational attraction will discover that these motions are governed by the same equations, whatever the velocity of the observers. But the "Galilean" transformation that relates the coordinates used by a moving observer to those of an observer at rest, and that leaves the equations of motion unchanged, would not leave the speed of light unchanged. To implement his new principle of relativity, Einstein needed to replace the Galilean coordinate transformation with a new sort of transformation, known as a Lorentz transformation, in which the motion of the observer affects not only the spatial coordinates of events but also their time coordinates. But if the equations of mechanics as well as the speed of light were to be invariant under such Lorentz transformations, then these equations too would now have to be changed. One of the most dramatic consequences of the new mechanics, pointed out by Einstein in 1907, is that a system that loses energy will experience a proportionate decrease in mass, given by the energy loss divided by the square of the speed of light. The energy available in a radioactive transmutation of one atomic nucleus into another could thus be determined by simply weighing the initial and final nuclei.

After 1905 only the theory of gravitation appeared inconsistent with Einstein's version of the principle of relativity. Einstein worked on this problem for the next decade, and finally found the answer in an extension of the principle of relativity. According to what he now called the General Principle of Relativity, the laws of nature appear the same not only to all uniformly moving observers, but (in a sense) to *all* observers: the familiar inertial forces that tell an observer that he is accelerating can alternatively be interpreted as a species of gravitation, caused by the acceleration of everything else relative to the observer. Since gravitational forces of any sort are just an aspect of the observer's coordinate system, they can be described geometrically, in terms of the curvature of spacetime, a curvature produced, according to Einstein's field equations, by the presence of energy and momentum.

Special and General Relativity, like quantum mechanics, contradict our everyday intuitive notions of physical reality. But the public reception of relativity and quantum mechanics was very different. Quantum mechanics was immediately confirmed by a wealth of preexisting data on atomic spectra, so there was no need for a dramatic critical experiment to prove it right. This, together with its greater conceptual difficulty, reduced its immediate impact on general culture. In contrast, there was little evidence at first for relativity, and particularly for General Relativity. Then, after World War I, a British expedition set out to measure an effect predicted by General Relativity, a tiny deflection of light from a distant star by the gravitational field of the sun. The news in 1919 that observations of this effect agreed with Einstein's prediction made headlines around the world, and raised Einstein to a unique public status among the scientists of the twentieth century.

Even though relativity had been much on the minds of de Broglie and Schrödinger, the successful initial applications of quantum mechanics in 1925–26 were based on the older mechanics of Newton. Electrons in their orbits in atoms travel at only a few percent at most of the speed of light, so the effects of relativity on atomic spectra are difficult to detect. But it was apparent that relativistic mechanics (like everything else) would have to be recast in a quantum framework.

This was not easy. In Special Relativity the observed flow of time is affected by the speed of the observer, so that even the order of events in time may vary from one observer to another. Nevertheless, classical Special Relativity respects the condition that all observers should see effects following their causes, because the order of events can depend on the speed of the observer only when the events are so close in time that light would not have time to travel from one event to the other, and Einstein assumed (for this reason) that no physical influence can travel faster than the speed of light. The problem of causality is more troublesome in quantum mechanics, because the positions and velocities in quantum states cannot both be specified with sufficient precision to rule out the propagation of physical effects at speeds faster than light.

The near incompatibility of Special Relativity with quantum mechanics was to have a profound effect on the style of theoretical physics. Both pre-quantum relativity and nonrelativistic quantum mechanics had provided general frameworks for physical theories, but physicists always had to rely on experiment for information about particles and forces in fashioning any specific theory. Relativistic quantum mechanics, in contrast, is so nearly internally inconsistent that a physicist can go far in describing nature by demanding mathematical consistency, with little input from observation.

Theory in this new style began with the work of Dirac in 1928. His approach was to revise the Schrödinger equation for a single particle moving in a given electromagnetic field to make it consistent with the Special Principle of Relativity. On this basis Dirac was able to calculate the strength of the electron's magnetic field and fine details of the hydrogen spectrum. More dramatically, several years later he predicted the existence of antimatter, a prediction unexpectedly confirmed with the 1932 discoveries of the antielectron in cosmic rays by Carl Anderson and by P. M. S. Blackett.

Dirac's approach gained great prestige from these successes, but it proved inadequate to deal with the radioactive process (known as beta decay) in which particles change their identity. The need to reconcile quantum mechanics and relativity soon gave rise to a more general formalism, that of the quantum theory of fields.

Quantum field theory began in 1926 with the application (by Born, Heisenberg, and Pascual Jordan) of quantum mechanics to the familiar fields of electromagnetism, which put the photon theory of light on a sound foundation. But electrons were still pictured as point particles, not as fields. Then in 1929–30 Heisenberg and Pauli proposed a more unified view of nature. Just as there is an electromagnetic field, whose energy and momentum come in tiny bundles called photons, so there is also an electron field, whose energy and momentum and electric charge are found in the bundles we call electrons, and likewise for every species of elementary particle. The basic ingredients of nature are fields; particles are derivative phenomena.

Quantum field theory reproduced the successes of Dirac's theory, and had some new successes of its own, but the synthesis of quantum mechanics with relativity soon ran into a new difficulty. In 1930 J. Robert Oppenheimer and Ivar Waller independently discovered that processes in which electrons emit and reabsorb photons would apparently shift the energy of the electron state by an infinite amount. Infinities soon turned up in other calculations, and produced a mood of pessimism about the validity of quantum field theory that persisted for decades.

The forefront of physics shifted in the 1930s from atoms to atomic nuclei, and to new phenomena revealed by the study of cosmic rays. Here are some highlights:

1932 JAMES CHADWICK discovers the neutron, an electrically neutral particle with a mass close to the proton's. Heisenberg proposes that the neutron is not a bound state of an electron and a proton, as was thought at first, but a new elementary particle, a constituent along with protons of atomic nuclei. Because they are electrically neutral, neutrons can penetrate atomic nuclei without being repelled by the strong electric fields near the nucleus, so they become a valuable tool for exploring the nucleus, especially in the hands of Enrico Fermi.

1933 FERMI develops a successful quantum field theory of beta decay (the radioactive process discovered by Becquerel). It describes how neutrons spontaneously change into protons and emit electrons and neutrinos (light neutral particles that had been proposed in 1930 by Pauli).

1935 HIDEKI YUKAWA offers a quantum field theory intended to account for both beta decay and for the strong nuclear forces that hold protons and neutrons together inside the nucleus. This theory requires the existence of a new particle, a "meson," with a mass of about 200 electron masses.

1937 S. H. NEDDERMEYER and Anderson, and C. E. Stevenson and J. C. Street, independently discover a new particle in cosmic rays. It has a mass close to that predicted by Yukawa, and is at first confused with his meson.

1939 HANS BETHE lays out the primary nuclear reactions by which stars gain their energy.

Basic research was in abeyance during World War II, while the discoveries of science were applied to war work. Applied research is outside the scope of this essay, but it is hardly possible to ignore entirely the most spectacular application of twentieth-century physics, the development of nuclear weapons. In 1938 Otto Hahn and Fritz Strassmann discovered that isotopes of the medium-weight element barium were produced when the heavy nuclei of uranium were irradiated with neutrons. This was explained a few months later in Sweden by Otto Frisch and Lise Meitner. They calculated that the capture of a neutron could cause a heavy nucleus like uranium to break into medium-sized pieces, like barium nuclei, releasing some 200 million electron volts of energy. In itself, this would be just another nuclear reaction, not really so different from hundreds of previously studied reactions. The thing that made neutron-induced fission so important was the prospect that, along with the barium and other nuclei, several neutrons would be released in each fission, each of which might trigger another fission, leading to an explosive chain reaction. It would be like a forest fire, in which the heat from each burning tree helps to ignite more trees, but vastly more destructive.[1]

---

1. This is a good place to dispel the common misapprehension that nuclear weapons are somehow an outgrowth of Einstein's Special Theory of Relativity. As originally derived in 1907, Einstein's famous equation $E = mc^2$ says that the liberation of an energy E by any system will cause it to lose a mass m equal to E

As war came closer, most of the great powers began to study the possibility of using uranium fission as a weapon or a source of power. Britain took an early lead in this work, which was then transferred to the United States. In December 1942 a group under Enrico Fermi at the University of Chicago succeeded in producing a chain reaction in a uranium pile. The fission took place not in the common isotope U^{238}, but in a rare isotope, U^{235}, that makes up only 0.723 percent of the natural uranium used by Fermi, but that, unlike U^{238}, can be made to fission even by very slow neutrons. Another nuclide, the isotope Pu^{239} of plutonium, has the same property of easy fissionability, and though it is absent in nature, it can be produced in uranium piles.

The problem then was either to isolate enough U^{235} or to produce enough Pu^{239} to make a bomb, and to develop a method of setting it off. These problems were successfully attacked by a team of scientists and engineers headed by Oppenheimer. In August 1945 Hiroshima and Nagasaki were largely destroyed by U^{235} and Pu^{239} bombs, respectively, leading soon after to the surrender of Japan and the end of the war.

It would take me too far from my subject to comment on the effect of nuclear weapons on world affairs, but it may be appropriate to say a word about the effect of nuclear weapons on physics itself. The achievement of the bomb project in helping to end the war gave many of its physicists a natural feeling of pride, often tempered with a sense of responsibility for the damage done in Hiroshima and Nagasaki, and for the danger that nuclear weapons now posed to humanity. Some physicists devoted themselves from then on to working for arms control and nuclear disarmament.

---

divided by c^2, the square of the speed of light. This is just as true of a burning tree as of a nuclear bomb, except that the energy released in ordinary burning is too small to allow a measurement of the decrease in mass of the products of combustion. If one insists on saying that mass is converted to energy in nuclear reactions, then one must say the same of ordinary burning. But the true source of the energy released when a tree burns is the sunlight absorbed during the tree's life, and the true source of the energy that is released when a uranium nucleus fissions is the energy stored in the nucleus when it was formed in a supernova explosion, long before the earth condensed from the interstellar medium.

Politicians and other nonscientists now tended to look at physics as a source of military and economic power, and for decades gave unprecedented support to basic as well as applied physical research. At the same time, the image of the Hiroshima mushroom cloud had a powerful effect on the attitude of many cultural leaders and other citizens toward science; in many cases their former indifference changed to outright hostility. Time has moderated all these effects, though at the century's end they are not entirely gone.

By 1947 physicists were ready to return to fundamental problems. The use of improved photographic emulsions soon revealed a menagerie of new particles in cosmic rays. One of them, the pi meson, could be identified with the particle predicted by Yukawa.

To discuss these and other new developments, a meeting was convened at Shelter Island, off the coast of Long Island, in June 1947. The high point of the meeting was a report by Willis Lamb. He presented convincing experimental evidence of a small difference in energy between two states of the hydrogen atom—states that, according to Dirac's version of relativistic quantum mechanics, should have precisely the same energy. Evidence was also presented from I. I. Rabi's group at Columbia showing that the electron's magnetic field is about a tenth of a percent stronger than had been calculated by Dirac. Effects of this sort would be produced by so-called radiative corrections, due to the emission and reabsorption of photons by electrons in atoms. In fact, the Lamb energy shift was a special case of the effect that Oppenheimer had been trying to calculate in 1930 when he first encountered the troublesome infinities.

The theorists at Shelter Island engaged in intensive discussions of how to carry out these calculations and, in particular, how to make sure that the answers would come out finite. One idea had already been widely discussed in the 1930s. Perhaps the masses and electric charges appearing in the field equations for the electron and other particle fields are themselves infinite, and the infinities encountered in calculations of radiative corrections just go to correct or "renormalize" these masses and charges, giving them their observed (and of course finite) values. Soon after Shelter Island the renormalization idea was used in successful calculations of the Lamb energy shift by Bethe, by J. Bruce French and Victor Weisskopf, and by Norman Kroll and Lamb, and the correction to the electron's magnetic field was calculated by Julian Schwinger. It was not known at the time in the West, but similar calculations were being carried out at the same time by a group in Japan headed by Sin-itiro Tomonaga.

These successes led to a renewed confidence in quantum field theory. Various radical ideas that had been put forward as solutions of the problem of infinities in the 1930s by Dirac, Heisenberg, and others were now set aside. But the method of calculation remained obscure and difficult, and it was hard to see whether renormalization would continue to resolve the problem of infinities in future calculations.

In the next few years after Shelter Island powerful methods of calculation were developed by Schwinger and Richard Feynman. The essence of these methods was the treatment of processes involving antiparticles in parallel with the corresponding processes involving particles, in such a way as to maintain explicit consistency with special relativity. Finally in 1949 the work of Feynman, Schwinger, and Tomonaga was synthesized by Freeman Dyson, putting quantum field theory in its modern form.

After 1950 the forefront of physics moved again, away from nuclear physics and radiative corrections, and toward the physics of a growing list of (supposedly) elementary particles. The tools of this research also changed; powerful accelerators supplanted radioactive nuclei and cosmic rays as a source of high energy particles, and particle detectors of increasing size and sophistication replaced the Geiger counters, cloud chambers, and photographic emulsions of the 1930s and 1940s. These technical improvements were accompanied by institutional changes, which, though probably inevitable, were not all welcome. Experimental research steadily moved from the basements of university physics buildings to large national or international laboratories, and physics articles appeared with increasing numbers of authors, some of them by the end of the century with a thousand authors from scores of institutions. The demarcation between theorists and experimenters became rigid; after the 1950s, no one again would do both experimental and theoretical work of high caliber in elementary particle physics.

The great task after 1950 was to bring all the known elementary particles and all the forces that act on them into the same quantum field-theoretic framework that had been used before 1950 to understand electrons and electromagnetic fields. The immediate obstacles were, first, that calculations in Fermi's theory of the weak nuclear forces responsible for beta decay revealed infinities that would not be removed by renormalization and, second, that the forces in Yukawa's theory of the strong nuclear forces are too strong to allow the kind of calculation by successive approximations that had worked so well in electrodynamics. Beyond these problems lay a deeper difficulty: there was no rationale for any of these theories.

By the mid-1970s these problems had been overcome, with the completion of a quantum field theory known as the Standard Model of elementary particles.

The Standard Model was the product of a century-long preoccupation with principles of symmetry in theoretical physics. In general, a symmetry principle is a statement that the equations of physics appear the same from certain different points of view. Symmetry principles had always been important in physics because they allow us to draw inferences about complicated systems without detailed knowledge of the system, but they attained a new importance in the twentieth century as aspects of nature's deepest laws.

Already in 1905 Einstein had elevated a principle of symmetry—invariance under Lorentz transformations of spacetime—to a status as a fundamental law of physics. Measurements of nuclear forces in the 1930s suggested a further symmetry—a rotation not in ordinary space or spacetime, but in an abstract internal "isotopic-spin" space, in which different directions correspond to different species of particles. One class of symmetries, known as gauge symmetries, actually requires the existence of fields, as, for instance, the General Principle of Relativity requires the existence of the gravitational field. Theorists from 1954 on proposed new gauge symmetries, though not yet with any clear application to the real world. But by the end of the 1950s experiments had revealed a bewildering variety of other, non-gauge, symmetry principles, many of them (like isotopic-spin symmetry, and the symmetry between left and right) only approximate.

Given the importance of symmetry principles, it was exciting to discover that nature respects yet other symmetries that are "broken"; they are not respected by individual processes or particles, and are manifested instead in relations between the probabilities for reactions involving different numbers of particles. Broken symmetry became a hot topic after the successful use of one of these broken symmetries in the mid-1960s to predict the properties of low energy pi mesons.

Soon after, a broken exact internal gauge symmetry was introduced as the basis of a unified theory of the weak and electromagnetic forces, which offered promise that the infinities that had plagued the old Fermi theory would be canceled by renormalization of the parameters of the theory. In 1971 it was proved that infinities in theories of this sort do indeed cancel in this way. The predictions of the new "electroweak" theory were dramatically confirmed in 1973, with the discovery of a new sort of weak force, and a decade later with the discovery of the predicted W and Z particles that carry these forces.

This left the problem of the strong nuclear forces. One valuable clue was the surprising discovery in 1968 that when probed by electrons of very short wavelength, neutrons and protons behave as if they are composed of particles that interact only weakly. In 1973 this phenomenon was clarified by a mathematical technique known as the "renormalization group," which had previously been applied in quantum electrodynamics and the theory of phase transitions. It turned out that in a particular quantum field theory known as quantum chromodynamics the strong forces do become weaker at very small scales. Quantum chromodynamics is a theory of strongly interacting particles known as quarks and gluons (the constituents of neutrons, protons, pi mesons, and so on) based on an exact unbroken internal gauge symmetry. This weakening of the strong forces made it feasible to calculate reaction rates by the same techniques earlier used in quantum electrodynamics. Experiments at high energy (and on several types of quarks of large mass, discovered starting in 1974) probed these small scales, where calculations are possible, and confirmed that quantum chromodynamics does indeed describe the strong forces.

The electroweak theory and quantum chromodynamics together constitute what is known as the Standard Model. The structure of the model is strongly constrained by its exact spacetime and internal symmetries, and by the need to avoid infinities. Both Fermi's theory of beta decay and Yukawa's theory of nuclear forces are explained in the standard model as low energy approximations. One happy by-product of the Standard Model, which contributed to its rapid acceptance, was that it also explained the known approximate symmetries as accidental consequences of the model's constrained structure.

The weakening of the strong forces at small scales makes it possible that the strong, weak, and electromagnetic forces should all become of the same strength at some very small scale. Calculations in 1974 showed that the strengths of these forces at accessible scales are consistent with this idea, and suggested that the scale at which the forces become of equal strength is about fifteen orders of magnitude smaller than the size of an atomic nucleus.

After the mid-1970s theoretical physics entered on a period of acute frustration. The Standard Model is clearly not the final answer; it incorporates some arbitrary features, and it leaves out gravitation. Another reduction is called for, one that would explain the Standard Model and General Relativity in terms of a simpler, more universal theory. Many theorists have tried to take this step, inventing attractive ideas of increasing mathematical sophistication—supersymmetry, supergravity, technicolor, string theory, and so on—but none of their efforts has been validated by experiment. At the same time, experimenters have continued to pile up evidence confirming the Standard Model, but, despite strong efforts, they have not discovered anything that would give theorists a clear clue to a deeper theory. Some help was expected from experiments that would clarify the one uncertainty left in the Standard Model—the detailed mechanism by which the electroweak gauge symmetry is broken—but a slowdown of research funding in the 1990s has put this off to the next century. It seems likely that a deeper, more unified theory will deal with structures at very small scales, perhaps sixteen to eighteen orders of magnitude smaller than an atomic nucleus, where all forces including gravitation may have similar strength. Unfortunately, these scales seem hopelessly beyond the range of direct experimental study. For physicists, the twentieth century seems to be ending sadly, but perhaps this is only the price we must pay for having already come so far.

20

A Designer Universe?

As part of its Program of Dialogue between Science and Religion, the American Association for the Advancement of Science held a symposium in April 1999 on the question of whether the universe shows signs of intelligent design. I was invited to speak, possibly as a representative nonbeliever; I was already known to be less than enthusiastic about religion, as shown for instance by the chapter "What about God?" in my book *Dreams of a Final Theory*.

One of the attractions in this invitation was the setting: the meeting would be held in the Baird Auditorium of the Smithsonian Institution's Museum of Natural History in Washington. This is the auditorium where in 1920 the astronomers Heber Curtis and Harlow Shapley held a famous debate on the size of the universe. Another attraction to me in this invitation was that I would be sharing the stage with an old friend, Sir John Polkinghorne.

When I first came to know Polkinghorne he was a talented mathematical physicist, working on the theory of strong nuclear forces. Sometime in the summer of 1975 or thereabouts, my wife and I were sitting in John's kitchen when he announced that he was going to give up his professorship at the University of Cambridge and take holy orders in the Church of England. I almost fell off my chair. I could imagine John and his wife, Ruth, following in the footsteps of David Livingstone, and preaching to the indigenes in one of the less comfortable parts of the world. As it happened, fate was kinder to him—he wound up as president of Queen's College, Cambridge, with a KBE. Over the years John had written several books on the relation between science and religion, with which I heartily disagreed.

At the meeting John outlined his views, and I outlined mine. We didn't get into much of a debate with each other. Nevertheless, the confrontation attracted a good deal of attention, with articles in *The Chronicle of Higher Education* and a dozen newspapers, including *The New York Times*. I submitted the edited transcript of my talk to Robert Silvers at *The New York Review of Books*, and he decided to publish it as a freestanding article, rather than as part of a debate with Polkinghorne.

This is the latest (and the most heart-felt) of my articles for *The New York Review of Books*. It has given rise to more comment—some quite hostile—than any other article I have written. My answers to the letters that were published in *The New York Review* are given in the next essay in this collection.

One frequent theme in other letters and articles responding to this essay is anger at one thing I said: "With or without religion, good people can behave well and bad people can do evil; but for good people to do evil—that takes religion." Several correspondents called my attention to the fact that the worst evils of the twentieth century were caused by regimes that had rejected religion: Hitler's Germany, Stalin's Soviet Union, Mao's China, and Pol Pot's Cambodia. This list leaves out some pretty horrible regimes in this century that enjoyed support from religious leaders—the Czar's Russia, Franco's Spain, Horthy's Hungary, the Ustashe's Croatia, the Taliban's Afghanistan, the Ayatollah's Iran, and so on. Even Hitler had the benefit of the 1933 concordat with the Catholic Church. But this is all off the point. Who would call Hitler, Stalin, Mao, Pol Pot, or their followers good men? In saying that it takes religion for good men to do evil I had in mind someone like Louis IX. By all accounts he was modest, generous, and concerned to an unusual degree with the welfare of the common people of France, but he was led by his religion to launch the war of aggression against Egypt that we know as the Sixth Crusade.

I never claimed that religion causes all the evil in the world, but I have learned that when you say anything controversial, you are likely to be blamed not so much for what you have said as for what people think that someone who has said what you said would also say. Still, it's better than finding that you have made no one angry.

But not everyone was angry at this article; it was chosen for inclusion in two different collections of the best essays of the year: *The Best American Essays, 2000* and *The Best American Science Writing, 2000.*

I have been asked to comment on whether the universe shows signs of having been designed. I don't see how it's possible to talk about this without having at least some vague idea of what a designer would be like. Any possible universe could be explained as the work of some sort of designer. Even a universe that is completely chaotic, without any laws or regularities at all, could be supposed to have been designed by an idiot.

The question that seems to me to be worth answering, and perhaps not impossible to answer, is whether the universe shows signs of having been designed by a deity more or less like those of traditional monotheistic religions—not necessarily a figure from the ceiling of the Sistine Chapel, but at least some sort of personality, some intelligence, who created the universe and has some special concern with life, in particular with human life. I suspect that this is not the idea of a designer held by many here. You may tell me that you are thinking of something much more abstract, some cosmic spirit of order and harmony, as Einstein did. You are certainly free to think that way, but then I don't know why you use words like "designer" or "God," except perhaps as a form of protective coloration.

It used to be obvious that the world was designed by some sort of intelligence. What else could account for fire and rain and lightning and earthquakes? Above all, the wonderful abilities of living things seemed to point to a creator who had a special interest in life. Today we understand most of these things in terms of physical forces acting under impersonal laws. We don't yet know the most fundamental laws, and we can't work out all the consequences of the laws we do know. The human mind remains extraordinarily difficult to understand, but so is the weather. We can't predict whether it will rain one month from today, but we do know the rules that govern the rain, even though we can't always calculate their consequences. I see nothing about the human mind any more than about the weather that stands out as beyond the hope of our understanding it as a consequence of impersonal laws acting over billions of years.

There do not seem to be any exceptions to this natural order, any miracles. I have the impression that these days most theologians are embarrassed by talk of miracles, but the great monotheistic faiths are founded on miracle stories—the burning bush, the empty tomb, an angel dictating the Koran to Mohammed—and some of these faiths teach that miracles continue at the present day. The evidence for all these miracles seems to me to be considerably weaker than the evidence for cold fusion, and I don't believe in cold fusion. Above all, today we understand that even human beings are the result of natural selection acting over millions of years.

I'd guess that if we were to see the hand of the designer anywhere, it would be in the fundamental principles, the final laws of nature, the book of rules that govern all natural phenomena. We don't know the final laws yet, but as far as we have been able to see, they are utterly impersonal and quite without any special role for life. There is no life force. As Richard Feynman has said, when you look at the universe and understand its laws, "the theory that it is all arranged as a stage for God to watch man's struggle for good and evil seems inadequate."

True, when quantum mechanics was new, some physicists thought that it put humans back into the picture, because the principles of quantum mechanics tell us how to calculate the probabilities of various results that might be found by a human observer. But, starting with the work of Hugh Everett forty years ago, the tendency of physicists who think deeply about these things has been to reformulate quantum mechanics in an entirely objective way, with observers treated just like everything else. I don't know if this program has been completely successful yet, but I think it will be.

I have to admit that, even when physicists will have gone as far as they can go, when we have a final theory, we will not have a completely satisfying picture of the world, because we will still be left with the question "why?" Why this theory, rather than some other theory? For example, why is the world described by quantum mechanics? Quantum mechanics is the one part of our present physics that is likely to survive intact in any future theory, but there is nothing logically inevitable about quantum mechanics; I can imagine a universe governed by Newtonian mechanics instead. So there seems to be an irreducible mystery that science will not eliminate.

But religious theories of design have the same problem. Either you mean something definite by a God, a designer, or you don't. If you don't, then what are we talking about? If you do mean something definite by "God" or "design," if for instance you believe in a God who is jealous, or loving, or intelligent, or whimsical, then you still must confront the question "why?" A religion may assert that the universe is governed by that sort of God, rather than some other sort of God, and it may offer evidence for this belief, but it cannot explain why this should be so.

In this respect, it seems to me that physics is in a better position to give us a partly satisfying explanation of the world than religion can ever be, because although physicists won't be able to explain why the laws of nature are what they are and not something completely different, at least we may be able to explain why they are not *slightly* different. For instance, no one has been able to think of a logically consistent alternative to quantum mechanics that is only slightly different. Once you start trying to make small changes in quantum mechanics, you get into theories with negative probabilities or other logical absurdities. When you combine quantum mechanics with relativity you increase its logical fragility. You find that unless you arrange the theory in just the right way you get nonsense, like effects preceding causes, or infinite probabilities. Religious theories, on the other hand, seem to be infinitely flexible, with nothing to prevent the invention of deities of any conceivable sort.

Now, it doesn't settle the matter for me to say that we cannot see the hand of a designer in what we know about the fundamental principles of science. It might be that although these principles do not refer explicitly to life, much less human life, they are nevertheless craftily designed to bring it about.

Some physicists have argued that certain constants of nature have values that seem to have been mysteriously fine-tuned to just the values that allow for the possibility of life, in a way that could only be explained by the intervention of a designer with some special concern for life. I am not impressed with these supposed instances of fine-tuning. For instance, one of the most frequently quoted examples of fine-tuning has to do with a property of the nucleus of the carbon atom. The matter left over from the first few minutes of the universe was almost entirely hydrogen and helium, with virtually none of the heavier elements like carbon, nitrogen, and oxygen that seem to be necessary for life. The heavy elements that we find on earth were built up hundreds of millions of years later in a first generation of stars, and then spewed out into the interstellar gas out of which our solar system eventually formed.

The first step in the sequence of nuclear reactions that created the heavy elements in early stars is usually the formation of a carbon nucleus out of three helium nuclei. There is a negligible chance of producing a carbon nucleus in its normal state (the state of lowest energy) in collisions of three helium nuclei, but it would be possible to produce appreciable amounts of carbon in stars if the carbon nucleus could exist in a radioactive state with an energy roughly 7 million electron volts (MeV) above the energy of the normal state, matching the energy in the mass of three helium nuclei, but (for reasons I'll come to presently) not more than 7.7 MeV above the normal state.

This radioactive state of a carbon nucleus could be easily formed in stars from three helium nuclei. After that, there would be no problem in producing ordinary carbon; the carbon nucleus in its radioactive state would spontaneously emit light and turn into carbon in its normal nonradioactive state, the state found on earth. The critical point in producing carbon is the existence of a radioactive state of the carbon nucleus that can be produced in collisions of three helium nuclei.

In fact, the carbon nucleus is known experimentally to have just such a radioactive state, with an energy 7.65 MeV above the normal state. At first sight this may seem like a pretty close call; the energy of this radioactive state of carbon misses being too high to allow the formation of carbon (and hence of us) by only 0.05 MeV, which is less than 1 percent of 7.65 MeV. It may appear that the constants of nature on which the properties of all nuclei depend have been carefully fine-tuned to make life possible.

Looked at more closely, the fine-tuning of the constants of nature here does not seem so fine. We have to consider the reason why the formation of carbon in stars requires the existence of a radioactive state of carbon with an energy not more than 7.7 MeV above the energy of the normal state. The reason is that the carbon nuclei in this state are actually formed in a two-step process: first, two helium nuclei combine to form the unstable nucleus of a beryllium isotope, beryllium 8, which occasionally, before it falls apart, captures another helium nucleus, forming a carbon nucleus in its radioactive state, which then decays into normal carbon. The total energy of the beryllium 8 nucleus and a helium nucleus at rest is 7.4 MeV above the energy of the normal state of the carbon nucleus; so if the energy of the radioactive state of carbon were more than 7.7 MeV it could be formed only in a collision of a helium nucleus and a beryllium 8 nucleus if the energy of motion of these two nuclei were at least 0.3 MeV—an energy which is extremely unlikely at the temperatures found in stars.

Thus the crucial thing that affects the production of carbon in stars is not the 7.65 MeV energy of the radioactive state of carbon above its normal state, but the 0.25 MeV energy of the radioactive state, an unstable composite of a beryllium 8 nucleus and a helium nucleus, above the energy of those nuclei at rest.[1] This energy

1. This was pointed out in a 1989 paper by M. Livio, D. Hollowell, A. Weiss, and J. W. Truran, "The Anthropic Significance of the Existence of an Excited State of ^{12}C," *Nature*, 340, no. 6231 (July 27, 1989). They did the calculation

misses being too high for the production of carbon by a fractional amount of 0.05 MeV/0.25 MeV, or 20 percent, which is not such a close call after all.

This conclusion about the lessons to be learned from carbon synthesis is somewhat controversial.[2] In any case, there *is* one constant whose value does seem remarkably well adjusted in our favor. It is the energy density of empty space, also known as the cosmological constant. It could have any value, but from first principles one would guess that this constant should be very large, and could be positive or negative. If large and positive, the cosmological constant would act as a repulsive force that increases with distance, a force that would prevent matter from clumping together in the early universe, the process that was the first step in forming galaxies and stars and planets and people. If large and negative, the cosmological constant would act as an attractive force increasing with distance, a force that would almost immediately reverse the expansion of the universe and cause it to recollapse, leaving no time for the evolution of life. In fact, astronomical observations show that the cosmological constant is quite small, very much smaller than would have been guessed from first principles.

It is still too early to tell whether there is some fundamental principle that can explain why the cosmological constant must be this small. But even if there is no such principle, recent developments in cosmology offer the possibility of an explanation of why the measured values of the cosmological constant and other physical constants are favorable for the appearance of intelligent life. According to the "chaotic inflation" and "eternal inflation" theories of André Linde, Alex Vilenkin, and others, the expanding cloud of billions of galaxies that we call the Big Bang may be just one fragment of a much larger universe in which big bangs go off all the time, each one with different values for the fundamental constants.

------

quoted here of the 7.7 MeV maximum energy of the radioactive state of carbon, above which little carbon is formed in stars.

2. H. Oberhummer, R. Pichler, and A. Csótó, Eötvös University preprint nucl-th/9810057 (1998).

In any such picture, in which the universe contains many parts with different values for what we usually call the constants of nature, there would be no difficulty in understanding why these "constants" take values favorable to intelligent life. There would be a vast number of big bangs in which the constants of nature take values unfavorable for life, and many fewer where life is possible. You don't have to invoke a benevolent designer to explain why we are in one of the parts of the universe where life is possible: in all the other parts of the universe there is no one to raise the question.[3] If any theory of this general type turns out to be correct, then to conclude that the constants of nature have been fine-tuned by a benevolent designer would be like saying, "Isn't it wonderful that God put us here on earth, where there's water and air and the surface gravity and temperature are so comfortable, rather than some horrid place, like Mercury or Pluto?" Where else in the solar system other than on earth could we have evolved?

Reasoning like this is called "anthropic." Sometimes it just amounts to an assertion that the laws of nature are what they are so that we can exist, without further explanation. This seems to me to be little more than mystical mumbo jumbo. On the other hand, if there really are a large number of worlds in which some constants take different values, then the anthropic explanation of why in our world they take values favorable for life is just common sense, like explaining why we live on the earth rather than Mercury or Pluto. The actual value of the cosmological constant, recently measured by observations of the motion of distant supernovas, is about what you would expect from this sort of argument:

3. The same conclusion may be reached in a more subtle way when quantum mechanics is applied to the whole universe. Through a reinterpretation of earlier work by Stephen Hawking, Sidney Coleman has shown how quantum mechanical effects can lead to a split of the history of the universe (more precisely, in what is called the wave function of the universe) into a huge number of separate possibilities, each one corresponding to a different set of fundamental constants. See Sidney Coleman, "Black Holes as Red Herrings: Topological Fluctuations and the Loss of Quantum Coherence," *Nuclear Physics,* B307 (1988), p. 867.

it is just about small enough so that it does not interfere much with the formation of galaxies. But we don't yet know enough about physics to tell whether there are different parts of the universe in which what are usually called the constants of physics really do take different values. This is not a hopeless question; we will probably be able to answer it when we have a good understanding of the quantum theory of gravitation.

It would be evidence for a benevolent designer if life were better than could be expected on other grounds. To judge this, we should keep in mind that a certain capacity for pleasure would readily have evolved through natural selection, as an incentive to animals who need to eat and breed in order to pass on their genes. It may not be likely that natural selection on any one planet would produce creatures who are fortunate enough to have the leisure and the ability to do science and think abstractly, but our sample of what is produced by evolution is very biased, by the fact that it is only in these fortunate cases that there is anyone thinking about cosmic design. Astronomers call this a selection effect.

The universe is very large, and perhaps infinite, so it should be no surprise that, among the enormous number of planets that may support only unintelligent life and the still vaster number that cannot support life at all, there is some tiny fraction on which there are living beings who are capable of thinking about the universe, as we are doing here. A journalist who has been assigned to interview lottery winners may come to feel that some special providence has been at work on their behalf, but he should keep in mind the much larger number of lottery players whom he is not interviewing because they haven't won anything. Thus to judge whether our lives show evidence for a benevolent designer, we have not only to ask whether life is better than would be expected in any case from what we know about natural selection, but also to take into account the bias introduced by the fact that it is we who are thinking about the problem.

This is a question that you all will have to answer for yourselves. Being a physicist is no help with questions like this, so I have to speak from my own experience. My life has been remarkably happy, perhaps in the upper 99.99 percentile of human happiness, but even so, I have seen a mother die painfully of cancer, a father's personality destroyed by Alzheimer's disease, and scores of second and third cousins murdered in the Holocaust. Signs of a benevolent designer are pretty well hidden.

The prevalence of evil and misery has always bothered those who believe in a benevolent and omnipotent God. Sometimes God is excused by pointing to the need for free will. Milton gives God this argument in *Paradise Lost:*

> I formed them free, and free they must remain
> Till they enthral themselves: I else must change
> Their nature, and revoke the high decree
> Unchangeable, eternal, which ordained
> Their freedom; they themselves ordained their fall.

It seems a bit unfair to my relatives to be murdered in order to provide an opportunity for free will for Germans, but even putting that aside, how does free will account for cancer? Is it an opportunity of free will for tumors?

I don't need to argue here that the evil in the world proves that the universe is not designed, but only to argue that there are no signs of benevolence that might have shown the hand of a designer. But in fact the perception that God cannot be benevolent is very old. Plays by Aeschylus and Euripides make a quite explicit statement that the gods are selfish and cruel, though they expect better behavior from humans. God in the Old Testament tells us to bash the heads of infidels and demands of us that we be willing to sacrifice our children's lives at His orders, and the God of traditional Christianity and Islam damns us for eternity if we do not worship Him in the right manner. Is this a nice way to behave? I know, I know, we are not supposed to judge God according to human standards, but you see the problem here: If we are not yet convinced of His existence, and are looking for signs of His benevolence, then what other standards *can* we use?

The issues that I have been asked to address here will seem to many to be terribly old-fashioned. The "argument from design" made by the English theologian William Paley is not on most peoples' minds these days. The prestige of religion seems today to derive from what people take to be its moral influence, rather than from what they may think has been its success in accounting for what we see in nature. Conversely, I have to admit that, although I really don't believe in a cosmic designer, the reason that I am taking the trouble to argue about it is that I think that on balance the moral influence of religion has been awful.

This is much too big a question to be settled here. On one side, I could point out endless examples of the harm done by religious enthusiasm, through a long history of pogroms, crusades, and jihads. In our own century it was a Muslim zealot who killed Sadat and a Jewish zealot who killed Rabin. No one would say that Hitler was a Christian zealot, but it is hard to imagine Nazism taking the form it did without the foundation provided by centuries of Christian anti-Semitism. On the other side, many admirers of religion would set countless examples of the good done by religion. For instance, in his recent book *Imagined Worlds,* the distinguished physicist Freeman Dyson has emphasized the role of religious belief in the suppression of slavery. I'd like to comment briefly on this point, not to try to prove anything with one example, but just to illustrate what I think about the moral influence of religion.

It is certainly true that the campaign against slavery and the slave trade was greatly strengthened by devout Christians, including the Evangelical layman William Wilberforce in England and the Unitarian minister William Ellery Channing in America. But Christianity, like other great world religions, lived comfortably with slavery for many centuries, and slavery was endorsed in the New Testament. So what was different for antislavery Christians like Wilberforce and Channing? There had been no discovery of new sacred scriptures, and neither Wilberforce nor Channing claimed to have received any supernatural revelations. Rather, the eighteenth century had seen a widespread increase in rationality and humanitarianism that led others—for instance, Adam Smith, Jeremy Bentham, and Richard Brinsley Sheridan—also to oppose slavery, on grounds having nothing to do with religion. Lord Mansfield, the author of the decision in Somersett's Case, which ended the legal protection of slavery in England (though not its colonies), was no more than conventionally religious, and his decision did not mention religious arguments. Although Wilberforce was the instigator of the campaign against the slave trade in the 1790s, this movement had essential support from many in Parliament like Charles James Fox and William Pitt, who were not known for their piety. As far as I can tell, the moral tone of religion benefited more from the spirit of the times than the spirit of the times benefited from religion.

Where religion did make a difference, it was more in support of slavery than in opposition to it. Arguments from scripture were used in Parliament to defend the slave trade. Frederick Douglass told in his *Narrative* how his condition as a slave became worse when his master underwent a religious conversion that allowed him to justify slavery as the punishment of the children of Ham. Mark Twain described his mother as a genuinely good person, whose soft heart pitied even Satan, but who had no doubt about the legitimacy of slavery, because in years of living in antebellum Missouri she had never heard any sermon opposing slavery, but only countless sermons preaching that slavery was God's will. With or without religion, good people can behave well and bad people can do evil; but for good people to do evil—that takes religion.

In an e-mail message from the American Association for the Advancement of Science I learned that the aim of this conference is to have a constructive dialogue between science and religion. I am all in favor of a dialogue between science and religion, but not a constructive dialogue. One of the great achievements of science has been, if not to make it impossible for intelligent people to be religious, then at least to make it possible for them not to be religious. We should not retreat from this accomplishment.

21

"A Designer Universe?":
An Exchange

Three of the letters responding to the foregoing essay were selected for publication by *The New York Review of Books*.

Professor Edward T. Oakes, S.J., of the Department of Religious Studies at Regis University, was good enough to say that I had raised weighty questions, but he remarked that "Mr. Weinberg shows a fundamental misunderstanding of the believer's position" when I applied the "why" question to God. He argued that for believers "the why-question takes on a whole new dimension when applied to God, who by definition is the ground that makes possible all why-questions whatever." Professor Regis also quoted a remark of Wittgenstein: "It isn't sensible to be furious even at Hitler; how much less so at God."

Professor Steven Goldberg of the CCNY Department of Sociology raised the question whether the argument against intelligent design of the universe is effectively unfalsifiable, in the sense of Karl Popper—something that could never be proved wrong, and hence can't be considered part of science. The same question had been raised after the talk in the Museum of Natural History on which this article was based. I answered there that, far from presenting an unfalsifiable view, I was actually going out on a limb. For instance, I said, a flaming sword might come down and strike me dead on the stage, which would provide pretty good evidence for divine intervention in the working of the universe. John Polkinghorne kindly interposed that he hoped that would not happen. If it did, he added, it would pose a theological problem for him as well as for me, since he did not believe in a God who would do that. To lighten the exchange, I pointed out that it would pose not only a theological problem, but also a janitorial problem.

The poet Anthony Hecht wrote to say that he hoped that I had misspoken myself when I declared that my life had been happy. He said that "as Solon would have informed him" the good things in my life could be set down to good fortune, and he argued that "happiness cannot easily abide alongside a consciousness of the misery and misfortunes of others." I didn't get this reference to Solon; I knew that he was an Athenian law-giver who had left no surviving written work. A classically trained friend, Paul Woodruff, directed me to the *Histories* of Herodotus, where I found the story of Solon's visit to Croesus, the king of Lydia. Croesus asks Solon to name the happiest man that he had ever met, and Solon lists several people he had known, none of whom were still alive. Croesus becomes vexed at this, and asks Solon why he considers these common people to be more happy than he, a king. Solon answers that life is so uncertain that until a man is dead, one should keep the word "happy" in reserve; until then one is merely fortunate.

My answers to these comments are given in the brief essay below. Since Mr. Hecht is a poet, I thought he could best be answered by a quote from another poet; the choice of the Yeats poem *Lapis Lazuli* was suggested by my wife.

I am staggered by Father Oakes's use of the definite article in his complaint that I show a fundamental misunderstanding of the believer's position. Surely there are many different positions taken by the many different people who assert the existence of God. Although it is not quite clear from his letter, I gather that the particular position taken by Father Oakes is that God "by definition" is the ultimate answer to any chain of "why" questions. I suppose I could have said that the laws of nature are by definition the ultimate answer to all "why" questions, but definitions can only take us so far. When we use the word "why" in a question—"Why was the morning newspaper missing?" or "Why does ice float on water?" or "Why do fools fall in love?"—we are expressing a sense of wonder, which can be allayed by an explanation—"The newspaper deliverer overslept," etc.—though that always leads to another "why" question. I confessed in my article that I cannot conceive of any set of laws of nature whose discovery would not leave me with an unsatisfied sense of wonder. In ruling out the application of the "why" question to the nature of God, Father Oakes in effect is claiming that he feels no sense of wonder about why the God in which he believes is the way He is. That may be so, but Father Oakes cannot have reached this happy state by any logical process—certainly not by definition.

On the remark of Wittgenstein quoted approvingly by Father Oakes, I must say that anyone, even Wittgenstein, who is not "furious" at Hitler should not be taken as a moral authority on any issue.

Professor Goldberg agrees with so much of what I said that I feel churlish in disagreeing with him about anything, but I think that in fact my nondesigner argument is eminently falsifiable. All that's needed is a miracle or two. In reply to a question after my talk at the American Association for the Advancement of Science, I suggested that everything I had said would be refuted if a flaming sword were to strike me down at the podium. There is also a less dramatic but more quantitative argument for a benevolent designer that might make sense if it had turned out that the earth on which we live were the only planet in the universe. Suppose that calculations showed that the chance of any single planet having a surface gravity, temperature, and chemical composition favorable for the appearance of life, and of life actually arising on this planet, and of this life becoming intelligent through natural selection, is no greater than one part in, say, a million million. Then with only one planet in the universe it would be difficult without supposing divine intervention to understand our great good fortune in having come into being. Of course, now we know that a good fraction of stars have planets, and that there are at least a hundred billion billion stars in the universe (perhaps an infinite number), so we need not be surprised that chance events governed by impersonal natural laws have produced intelligent life on at least one of the planets. With these odds, it would be surprising if intelligent life had not appeared.

Of all the comments on my article, the one I least expected was that of the poet Anthony Hecht. When I mentioned in my article that my life had been a happy one, it was because I was about to raise the issue of the pain of human life in general, and I did not want to seem to be whining about it. Mr. Hecht may be right that our knowledge of the misery and misfortunes of others should keep us from feeling happy, but that is not the position taken by Solon, whom Mr. Hecht invokes. In the story told by Herodotus, the reason that Solon gives for resisting the illusion of happiness is the riskiness of our own lives, not the misfortunes of others. To support our right to consider ourselves happy despite the warnings of both Solon and Mr. Hecht, I would call to witness the words of William Butler Yeats:

> All perform their tragic play,
> There struts Hamlet, there is Lear,
> That's Ophelia, that Cordelia;
> Yet they, should the last scene be there,
> The great stage curtain about to drop,
> If worthy their prominent part in the play,
> Do not break up their lines to weep.
> They know that Hamlet and Lear are gay;
> Gaiety transfiguring all that dread.

22

Five and a Half Utopias

In 1997 *The Atlantic Monthly* invited me to contribute an essay having something to do with the coming new millennium. Like many other people, I had been daydreaming about utopias of one sort and another, and I had never had a chance to write about this sort of thing, so I agreed to contribute an article about utopian ideas. I sent in the article, and the *Atlantic* editors accepted it, but then as time passed I heard nothing more about it. I had been paid for the article, so at first I didn't worry about it, but finally, early in 1999, I took another look at the article.

It was pretty awful. The first third was a review of every piece of utopian and anti-utopian literature I could remember, from Plato and More to Bellamy and Wells and Huxley and Orwell. It was not an article; it was a World Lit term paper, and a B- term paper at that. I gave it to Louise Weinberg to read, and she saw immediately what was needed: throw away the first third. I took her advice, rewrote the rest, and sent it to the *Atlantic*, which now moved quickly to get the article into the January 2000 issue.

There was something in the article for almost everyone to dislike, and the *Atlantic* received a fair number of angry letters. The two I liked best took me to task for saying that socialism has failed everywhere. One of these letters said that socialism has not failed because it has never been tried, while the other letter said that socialism has not failed because it has succeeded in taking us over, as shown by the income tax and Social Security. I wish that I could introduce these two letter writers to each other.

I used to read a good deal of science fiction when I was a boy. Even though I knew pretty early that I was going to be a scientist, it wasn't the science that interested me in science fiction; it was the vision of future societies that, for better or worse, would be radically different from our own. This led me on from science fiction to utopian literature, to Plato's *Republic*, Thomas More's *Utopia*, and Edward Bellamy's *Looking Backward*, and also to the literature of anti-utopias, to Aldous Huxley's *Brave New World* and George Orwell's *1984*. I have been more interested in other things in recent years, but now that we are starting a new millennium, it is natural to start thinking again about what sort of utopia or anti-utopia might be waiting for us in the future.

There was a great deal of this sort of speculation at the end of the previous century. The characters in Anton Chekhov's *Three Sisters* (written just one hundred years ago) seem captivated by utopian dreams. Here, for instance, is Colonel Vershinin, in Act II:

> In a century or two, or in a millennium, people will live in a new way, a happier way. We won't be there to see it—but it's why we live, why we work. It's why we suffer. We're creating it. That's the purpose of our existence. The only happiness we can know is to work toward that goal.

Vershinin's hopes have not worked out so well in this century. The most influential utopian idea of the nineteenth and twentieth centuries was socialism, which has failed everywhere. Under the banner of socialism Stalin's Soviet Union and Mao's China gave us not utopias but ghastly anti-utopias. It is ironic that in the heyday of utopian thinking, in the nineteenth century, Karl Marx himself sneered at utopian thought, and claimed to be guided instead by a science of history. Of course, there is no science of history, but that's almost beside the point. Even if we could decide that some type of government or economy was historically inevitable, as Marx believed communism to be, it would not follow that this would be something we would like. If Marx had been an honest utopian, and recognized his responsibility to describe the society he wanted to bring into being, it might have been clearer from the beginning that the effort would end in tyranny. Hitler's Germany, too, started with utopian rhetoric: socialism combined with a maniac vision of a master race.

Even so, I can't believe that we have seen the last of utopia-mongering. Indeed, five nonsocialist styles of utopia seem (in various combinations) to be emerging in public debate. We had better watch out for people selling these utopias; each of these visions abandons one or more of the grand causes—equality, liberty, and the quality of life and work—that motivated the best utopian ideas of the past.

The Free-Market Utopia

Government barriers to free enterprise disappear. Governments lose most of their functions, serving only to punish crimes, enforce contracts, and provide national defense. Freed of artificial restraints, the world becomes industrialized and prosperous.

This style of utopia has the advantage of not depending on any assumed improvements in human nature, but that doesn't mean we have to like it. If only for the sake of argument, let's say that *something* (productivity? gross national product? Pareto efficiency?) is maximized by free markets. Whatever it is, we still have to decide for ourselves whether this is what we want to be maximized.

One thing that is clearly not maximized by free markets is equality. I am talking not about that pale substitute for equality known as equality of opportunity but about equality itself. Whatever purposes may be served by rewarding the talented, I have never understood why untalented people *deserve* less of the world's good things than other people. It is hard to see how equality can be promoted, and a safety net provided for those who would otherwise fall out of the bottom of the economy, unless there is government interference in free markets.

Not everyone has put a high value on equality. Plato did not have much use for it, especially after the Athenian democracy condemned his hero, Socrates. He explained the rigid stratification of his Republic by comparing society to the human soul: the guardians are the rational part; the soldiers are the spirited part; and the peasants and artisans are the baser parts. I don't know whether he was more interested in the self as a metaphor for the state or the state as a metaphor for the self, but at any rate such silly analogies continued for two millennia to comfort the comfortable.

In the course of time the dream of equality grew to become an emotional driving force behind utopian thinking. When English peasants and artisans rebelled against feudalism in 1381, their slogan was the couplet preached by John Ball at Blackheath: "When Adam delved, and Eve span, who was then the gentleman?" The French Revolution adopted the goal of equality along with liberty and fraternity; Louis-Philippe-Joseph, duc d'Orléans, wishing to gain favor with the Jacobins, changed his name to Philippe-Egalité. (Neither his new name nor his vote for the execution of Louis XVI saved the duke from the Terror, and he joined the king and thousands of other Frenchmen in the equality of the guillotine.) The central aim of the socialists and anarchists of the nineteenth and twentieth centuries was to end the unequal distribution of wealth. Bellamy followed *Looking Backward* with a sequel titled simply *Equality*. It is a cruel joke of history that in the twentieth century the passion for equality has been used to justify communist states in which everyone was reduced to an equality of poverty. Everyone, that is, except for a small number of politicians and celebrities and their families, who alone had access to good housing, good food, and good medicine. Egalitarianism is perhaps the aspect of utopian thinking that has been most discredited by the failure of communism. These days anyone who urges a more equal distribution of wealth is likely to be charged with trying to revive the class struggle.

Of course, some inequality is inevitable. Everyone knows that only a few people can be concert violinists, factory managers, or major-league pitchers. In revolutionary France the ideal of equality soon gave way to the *carrière ouverte aux talents*. It was said that each soldier in Napoleon's army carried a marshal's baton in his knapsack, but no one expected that many soldiers would get to use it. For my part, I would fight against any proposal to be less selective in choosing graduate students and research associates for the physics department in which I work. But the inequalities of title and fame and authority that follow inexorably from inequalities of talent provide powerful spurs to ambition. Is it really necessary to add gross inequalities of wealth to these other incentives?

This issue cannot be judged on purely economic grounds. Economists tell us that inequality of compensation fulfills important economic functions: just as unequal prices for different foods help in allocating agricultural resources to produce what people want to eat, so unequal rewards for labor and for capital can help in directing people into jobs, and their money into investments, of the greatest economic value. The difference between these various inequalities is that *in themselves,* the relative prices of wheat and rye are of no importance; they only serve the economic function of helping to adjust production and resources. But whatever its economic effects, gross inequality in wealth is itself a social evil, which poisons life for millions.

Those who grew up in comfortable circumstances often have trouble understanding this. They call any effort to reduce inequality "the politics of envy." The best place for the well-to-do to get some feeling for the damage done by inequality may be American literature, perhaps because America led the world in making wealth the chief determinant of class. This damage is poignantly described in the novels of Theodore Dreiser, who grew up poor during the Gilded Age, when inequality of wealth in America was at its height. Or think of Willa Cather's story "Paul's Case." The hopeless longing of the boy Paul for the life of the rich drives him to give up his whole dreary life for a few days of luxury.

Another thing that is manifestly not maximized by free markets is civilization. By "civilization" I mean not just art museums and grand opera but the whole range of public and private goods that are there not merely to help keep us alive but to add quality to our lives. Everyone can make his or her own list; for me, civilization includes classical-music radio stations and the look of lovely old cities. It does not include telemarketing or Las Vegas. Civilization is elitist; only occasionally does it match the public taste, and for this reason it cannot prosper if not supported by individual sacrifices or government action, whether in the form of subsidy, regulation, or tax policy.

The aspect of civilization that concerns me professionally is basic scientific research, like the search for the fundamental laws of nature or for the origins of the universe or of life—research that cannot be justified by foreseeable economic benefits. Along with all the good things that have come from the opening of free-market economies in Eastern Europe, we have seen the devastation in those countries of scientific establishments that cannot turn a profit. In the United States the opening of the telephone industry to free-market forces has led to the almost complete dismantling of pure science at the Bell Laboratories, formerly among the world's leading private scientific-research facilities.

It might be worthwhile to let equality and civilization take their chances in the free market if in return we could expect that the withering of government would serve as a guarantee against oppression. But that is an illusion. For many Americans the danger of tyranny lies not in government but in employers or insurance companies or health-maintenance organizations, from which we need government to protect us. To say that any worker is free to escape an oppressive employer by getting a different job is about as realistic as to say that any citizen is free to escape an oppressive government by emigrating.

The Best-and-Brightest Utopia

Public affairs are put in the hands of an intelligent and well-educated class of leaders.

This was Plato's vision. In the *Republic* and other dialogues Plato described a hierarchical society of peasants and soldiers ruled over by a eugenically bred class of "guardians," and in *Critias* he imagined that this was the constitution of ancient Athens, before the war with Atlantis, nine thousand years earlier. In our own times Lee Kuan Yew, the Senior Minister of Singapore, has said that only an elite, consisting of the top 3 to 5 percent of a society, can deal effectively with public issues. The rulers of the "People's Republic" of China would probably agree, except that I suppose they would think that 3 percent is a gross overestimate. Even democratic countries such as France and Japan recruit their powerful bureaucracies from special educational institutions—the Grandes Ecoles and the University of Tokyo.

The claims of Lee Kuan Yew and others for the effectiveness of "Asian model" technocracies look pretty unconvincing after the East Asian economic downturn of the past few years. Even before that, Amartya Sen and other economists had argued that authoritarian governments do not generally perform better economically than democratic ones, and may in fact be more at risk of economic catastrophe. But rule by an elite has much worse drawbacks.

As Alexis de Tocqueville pointed out, even if government by an elite could be trusted to be efficient and public spirited, it would have the effect of making its citizens into children. And surely we should have learned by now that no such government can be trusted. Behind every Marcus Aurelius is a crazy relative like Commodus, waiting to take over.

There never has been a governing elite in any age that did not eventually come to give priority to its own interests. It doesn't help to choose the elite from some special segment of society. Attacking Marxism, the anarchist Mikhail Bakunin pointed out that it would be impossible to put workers at the head of government, because then they would cease to be workers and instead would become governors. In *Looking Backward*, Bellamy, like many other socialists, argued that labor unions would become unnecessary once the means of production were handed over to a national industrial army, because then the workers would own their own factories. This argument was not borne out by the experience of labor in the Soviet Union, to say the least. There is no reason to imagine that a ruling elite drawn from business leaders would do any better. H. G. Wells and other utopians have imagined putting public affairs in the hands of scientists, but I know my fellow scientists too well to be enthusiastic about this proposal. Most scientists would rather do their own research than govern anyone. I have known a number of academic physics departments in which faculty members actively compete for the privilege of *not* being department chairman. Anyway, I haven't seen any signs that scientists would be better than anyone else at running a country.

Power is not safe in the hands of any elite, but it is not safe in the hands of the people, either. To abandon all constraints on direct democracy is to submit minorities to the tyranny of the majority. If it were not for the interposition of an elite judiciary, the majority in many states might still be enforcing racial segregation, and at the very least would have introduced prayer sessions in the public schools. It is the majority that has favored state-imposed religious conformity in Algeria and Afghanistan and other Islamic countries.

So what is the solution? Whom can we trust to exercise government power? W. S. Gilbert proposed an admirably simple solution to this problem. In the Savoy opera *Utopia, Limited,* the King exercises all power but is in constant danger of being turned over to the Public Exploder by two Wise Men, who explain,

> Our duty is to spy
>   Upon our King's illicities,
> And keep a watchful eye
>   On all his eccentricities.
> If ever a trick he tries
>   That savours of rascality,
> At our decree he dies
>   Without the least formality.

We just have to get used to the fact that in the real world there is no solution, and we can't trust anyone. The best we can hope for is that power be widely diffused among many conflicting government and private institutions, any of which may be allies in opposing the encroachments of others—much as in the United States today.

The Religious Utopia

A religious revival sweeps the earth, reversing the secularization of society that began with the Enlightenment. Many countries follow the example of Iran, and accept religious leaders as their rulers. America returns to its historical roots as a Christian country. Scientific research and teaching are permitted only where they do not corrode religious belief.

It is hard to see why anyone would think that religion is a cure for the world's problems. People have been at each other's throats over differences in religion throughout history, a sad story that continues today in Northern Ireland, the Balkans, the Middle East, Sudan, and India. But even fighting over religion is not as bad as an imposed religious uniformity. Of all the elites that can oppress us, the most dangerous are those bearing the banner of religion. Their power is greater, because they can threaten punishment in the next world as well as in this, and their influence is more intrusive, because it reaches into matters that ought to be left to private choice, such as sexual practice and family life. In our own times we have had a taste of what utopias based on religious uniformity are like, in countries like Iran, Saudi Arabia, and Afghanistan, where the freedom of women is sharply limited, and holy war is preached to children.

Religious readers may object that the harm in all these cases is done by perversions of religion, not by religion itself. But religious wars and persecutions have been at the center of religious life throughout history. What has changed, that these now seem to some people in some parts of the world to be only perversions of true religious belief? Has there been a new supernatural revelation, or a discovery of lost sacred writings that put religious teachings in a new light? No—since the Enlightenment there has been instead a spread of rationality and humanitarianism that has in turn affected religious belief, leading to a wider spread of religious toleration. It is not that religion has improved our moral sense but that a purely secular improvement in our moral values has improved the way religion is practiced here and there. People ought to be religious or not religious according to whether they believe in the teachings of religion, not because of any illusion that religion raises the moral level of society.

The Green Utopia

The world turns away from industrialism and returns to a simpler style of life. Small communities grow their own food, build houses and furniture with their own hands, and use electricity only to the extent that they can generate it from sun, wind, or water.

This is the sort of utopia that appears most often in modern literature—for instance, in the science fiction of Ursula Le Guin. But modern writers tend to locate their utopias on other planets. No one has described a rural utopia here on earth better than William Morris did in 1890, in *News from Nowhere*. (His title, by the way, is an echo of More's *Utopia*, which might come from either the Greek *eu-* + *topos*, meaning "good place," or *ou-* + *topos*, meaning "no place." The second meaning was also picked up by Samuel Butler, in *Erewhon* (1872), which of course is "nowhere" spelled backward—except that it isn't, which shows how hard it is to be perfect.) In Morris's future England, Hammersmith and Kensington are again small villages; the national government has become unnecessary; and the Houses of Parliament are used to store manure. Morris gives a lovely description of the unpolluted countryside seen by his hero in a long rowing voyage from London to the upper reaches of the Thames. It is all very pretty, but some of us would miss urban London.

It is common for those who don't have to work hard to romanticize hard labor, especially agricultural labor. Shakespeare's Henry V imagines that no king can sleep as soundly as a peasant,

> Who with a body fill'd and vacant mind
> Gets him to rest, cramm'd with distressful bread;
> Never sees horrid night, the child of hell,
> But, like a lackey, from the rise to set
> Sweats in the eye of Phoebus and all night
> Sleeps in Elysium; next day after dawn,
> Doth rise and help Hyperion to his horse,
> And follows so the ever-running year,
> With profitable labour, to his grave.

I doubt that any real peasant would see farm work this way. In the words of Mel Brooks, "It's good to be the king."

Some utopians—like Wells, in *The World Set Free*—would like to restore the natural environment of the past while keeping the benefits of technology, by radically reducing the earth's population. This seems hard on all those who would be unable to enjoy utopia because they had not been born. Others, like Morris, imagine that a nontechnological utopia could support the same population as at present. I don't believe it, but even if I did, I would object to abandoning the technology that gives us heart defibrillators and elementary particle accelerators. In fact, Morris cheats. He refers to some sort of "force" that helps with necessary work that can't be done by hand; but how could something like this exist without an industrial establishment?

Hostility to technology also promotes hostility to science, which gets additional fuel from the discomfort produced by what science reveals about the world. In a speech at Independence Hall, in Philadelphia, on the Fourth of July in 1994, the Czech poet and statesman Václav Havel protested that "we are not at all just an accidental anomaly . . . we are mysteriously connected to the entire universe." He called for "a science that is new . . . postmodern." One of the items that Havel would like to include in this new science is the Gaia hypothesis, according to which the earth and the living things it supports form a single organism. If the Gaia hypothesis is any more than a poetic way of expressing the obvious fact that life and its environment act on each other, then it is mystical nonsense, but it has a nice Green tinge that Havel obviously likes. This business of picking out the comforting parts of science and condemning the rest is an old story. The people of future England in *News from Nowhere* engage in some sort of science, about which Morris says only that it is different from the "commercial" science of the nineteenth century. This is an amazing comment on the science of Charles Darwin and James Clerk Maxwell. One gets the impression that the work of science in Morris's utopia consists of collecting pretty rocks and butterflies.

The Technological Utopia

The development of information processing, robotics, synthetic materials, and biotechnology increases productive capacity so much that questions about the distribution of wealth become irrelevant. National borders also become irrelevant, as the whole world is connected by a web of fiber-optic cables.

There is a tendency to exaggerate the rate at which our lives will be changed by technology. We still have a whole year to go before 2001, but I doubt that Arthur C. Clarke's vision of commercial flights to the moon is going to come true by then. Individual technologies reach plateaus beyond which further improvement is not worthwhile. For instance, the experience of riding in commercial aircraft has not materially changed since the introduction of the Boeing 707, more than forty years ago. (The Concorde is an exception that proves the rule; it has never paid for the cost of its development.) Computer technology clearly has not yet reached its plateau, but it will—probably when the miniaturization of solid state devices runs into the limits imposed by the finite size of individual atoms. Successful technologies also tend to be self-limiting once they become available to the general population. I doubt that it is possible to cross Manhattan from the East River to the Hudson River faster by automobile today than it was by horse-drawn streetcar a century ago. The Internet is already beginning to show the effects of overcrowding. I tremble at the thought of two billion air-conditioners in a future China and India, each adding its own exhaust heat to the earth's atmosphere.

Still, however long it may take, new technologies will inevitably bring great changes to our lives. Far from leading us to utopia, some of these changes may well be frightening. Technology certainly gives us the power to wreck the environment in which we live. Also, I can't imagine anything more destructive of common feeling among the world's people than a new medical technology that would extend youth for decades but would be affordable only by the very rich.

Then there is the problem of what people would do with themselves if technology freed most of them from the necessity of work. As Sigmund Freud taught, our greatest needs are love and work. Work gives us a sense of identity and the dignity of earning our living, and it gives many of us our chief reason to get out of the house. In "The Machine Stops," E. M. Forster imagined a world of perfect comfort whose people are isolated from one another within an all-caring machine. Their lives are so appalling that the reader is glad when the story's title comes true.

Some utopians imagine that the problem of work will solve itself. Wells vaguely suggested that after technology had brought universal plenty, everyone would become an artist, and Bellamy thought that when workers retired at age forty-five, many of them would take up the arts or the sciences. I can't think of any better way to spread general misery. Even a lover of the arts can read only so much new literature, hear only so much new music, or look at only so many new paintings or sculptures, and in trying to choose the best of these everyone will tend to be drawn to the same works. Consequently, whatever joy they took in the work itself, the great majority of writers, composers, painters, and sculptors would spend their lives without having anyone else notice their work. The same would apply to scientists. By now it is impossible for a theoretical physicist to read all the papers even in some narrow subspecialty, so most articles on theoretical physics have little impact and are soon forgotten.

Morris excluded modern technology from his utopia not only because he was in love with the Middle Ages but also because he wanted to preserve work for people to do. Although modern technology has made work more unsatisfying for many, I think that Morris was wrong in supposing that this is inevitable. The mindless, repetitive quality that makes routine jobs on assembly lines so hateful is also just the thing that would allow them in the future to be done entirely by machines. Technology creates better jobs, from auto mechanic to astronaut. But there is no guarantee that the advance of technology will provide all people with work that they like to do, and in the short run it converts the badly employed into the unemployed.

One of the things that attracts some people to technological utopias is the prospect of a world unified by technology. In the utopia of Wells's *The World Set Free* all national boundaries are dissolved; there is a powerful world government, a single world language (English, of course), worldwide adoption of the metric system, and interconvertible currencies with fixed exchange rates. There is still a United States in Bellamy's *Looking Backward,* but its citizens look forward to eventual world unification.

Physicists (who invented the World Wide Web) already participate in an early version of world unification. For instance, throughout the world we share a typesetting code for mathematical symbols known as LaTeX, based on English. I recently did some work on the quantum theory of fields in collaboration with a Catalan physicist who was visiting Kyoto; we sent our equations back and forth between Texas and Japan by e-mail, in LaTeX.

I am not so sure that world unification is an unmixed blessing. It has the side effect of shrinking the psychological space in which we live. A few hundred years ago large areas of the map were blank, leaving the imagination free to fill them with strange peoples and animals. Even Queen Victoria, who, it is said, tried to taste every fruit grown in the British Empire, never had a chance to try a mango or a durian. Now we can fly anywhere, and we buy mangoes in our local supermarkets. This is not my idea of utopia. Wouldn't it be more exciting to eat a mango if it could be done nowhere but in India? What is the good of getting somewhere quickly if it is no different from the place one has left?

More is at stake here than just making travel fun again sometime in the future when everyone can afford it. Isolated by language differences and national boundaries, each of the world's cultures represents a precious link to the past and an opportunity for distinctive new artistic and intellectual creation. All these are put at risk by steps toward world unification.

Now I have said hard things about five different styles of utopia—so what do I have to offer? No easy solutions. There is no simple formula that will tell us how to strike a balance between the dangers from governing elites and those from majority rule or free markets, or between the opportunities and the hazards of new technology. I can't resist offering a utopian vision of my own, but it is a very modest one.

The Civilized Egalitarian Capitalist Utopia

Production remains mostly in the hands of competing private corporations, overseen by a democratic government that is itself overseen by independent courts; these corporations continue to use high salaries along with status and authority to attract workers and managers with special talents, and dividends to attract capital. Those who receive a high income are able to keep only part of it; to prevent the rest of their income from being simply taken in taxes they give much of it to museums, universities, and other institutions of their choice, reaping benefits that range from moral satisfaction to better seats at the opera. These nonprofit institutions use the donations to invest in business enterprises, eventually replacing wealthy individuals as the owners of industrial corporations.

Not very original? No, it is in fact a natural development from some present trends. Nonprofit institutions have been the fastest-growing sector of the American economy over the past fifteen years. But the tide of American politics now seems to be flowing in the opposite direction. We are in the process of giving up our best weapon against inequality: the graduated income tax, levied on all forms of income and supplemented by taxes on large legacies. A steeply graduated income tax, if accompanied by generous allowances for the deduction of charitable contributions, has another virtue: it amounts to a public subsidy for museums, symphony orchestras, hospitals, universities, research laboratories, and charities of all sorts, without putting them under the control of government. Oddly, the deductibility of charitable contributions has been attacked in whole or in part by conservatives like Steve Forbes and Herbert Stein, even though it has been a peculiarly American way of achieving government support for the values of civilization without increasing government power.

I don't offer this modest utopia with any great fervor, because I have doubts whether men and women will be content with an individualistic life of love and work and liberty and equality. People have seemed also to need some exciting collective enterprise that, even if destructive, would lift them out of the everyday round of civilized life.

The individualistic lives of propertied European men at the beginning of the twentieth century were about as pleasant as one can imagine: these men moved in a world of elegant cafés, theaters, country houses, and relatively unspoiled countryside; their comforts were seen to by deferential women and servants; and for those who cared about such things, there were exciting innovations in science and the arts. Yet there is plenty of evidence that many of these men were afflicted with such boredom and directionlessness that they felt as they went off to the Great War in 1914, like "swimmers into cleanness leaping." Now war has become intolerable. Perhaps someday we may find a better common cause in the colonization of the solar system, but that is far off, and even then most people will be left here on earth.

Can we change ourselves enough to be satisfied with a civilized society? The dream that behaviorists and Marxists had of changing human nature seems to me the worst sort of exaggeration of the capabilities of science. In *Three Sisters,* Chekhov has Baron Tuzenbach reply to Vershinin's utopian dreams:

> Well, maybe we'll fly in balloons, the cut of jackets will be different, we'll have discovered a sixth sense, maybe even developed it—I don't know. But life will be the same—difficult, full of unknowns, and happy. In a thousand years, just like today, people will sigh and say, oh, how hard it is to be alive. They'll still be scared of death, and won't want to die.

Facing a new millennium we can share some of Vershinin's hopes for utopia, but when it comes to judging the chances of really changing the way we live, no doubt most of us would side with Tuzenbach.

23

Looking for Peace
in the Science Wars

In May 1999 I received an e-mail message from Maren Meinhardt, an editor at London's *Times Literary Supplement*, asking if I would be interested in reviewing Ian Hacking's book *The Social Construction of What?* (Harvard University Press, 1999). I was already reading Hacking's book (as I told Meinhardt, I had seen several references to me in the index, which I always take as a good sign), and I had admired the *TLS* for many years, so I agreed to have a try at it. In writing the review I found that I was returning to the same arguments with the social constructivists that I had entered with two essays in this collection: "Night Thoughts of a Quantum Physicist" and "Sokal's Hoax." This review is probably the last time that I will get into this debate, as it is not likely that there is anyone left whose opinions on these issues are still susceptible to change.

"Why the devil came you between us?" said Mercutio. That is the risk that mediators have to run—like Romeo, they may only make the quarrel worse. Now the distinguished Canadian philosopher Ian Hacking is intervening in the quarrel between the academics (many of them sociologists) who talk about the "social construction" of this or that, and those others (many of them natural scientists) who deplore such talk. While he does not promise peace in our time, Hacking criticizes the way that scientists and sociologists shout at each other, and he laments talk of culture wars. I am not sure that he won't make the quarrel worse, but I admire the style and courage of his intervention.

What is the quarrel about? My own university library lists sixty-eight books with "the social construction" of something or other in the title: you can read about the social construction of gender, of anorexia, of disaster, of leisure for women in academe, and so on. Clearly these authors do not mean simply that the thing in question is partly or wholly produced by society. That is too obviously true for leisure, anorexia, and so on. Instead, they generally mean that the way we think about things and even the fact that we think that we *should* think about them are not just consequences of the way the world is, but are conditioned by our immersion in a particular society. As Hacking says, there is a kind of "unmasking" going on when academics write about social construction. "Silly you" (the subtext goes), "you thought that anorexia was just one of the diseases that flesh is heir to, but the same behavior in another society might not be recognized as a disease or even a particular pattern of behavior at all." Sometimes there is a further subtext, that the world would be a better place if we would not take seriously whatever it is that has been unmasked as a social construction.

Talking about social construction in this way may be wise or foolish, depending on what it is that is supposed to be socially constructed, though I cannot help feeling that the phrase is being overused by a whole lot of Ph.D. candidates. Hacking has some interesting things to say about the social construction of madness and child abuse, or, rather, as he emphasizes, of our *ideas of* madness and child abuse. But it is not talk about the social construction of ideas of human behavior that gets the fur flying in the faculty club. The way to get academics really worked up on one side or the other is to talk about the social construction of knowledge in the exact sciences: physics, chemistry, molecular biology.

It is not easy to keep track of the different positions taken by all the sociologists and historians who adopt a social constructivist view of science. At one time, there was a cluster of them at the

University of Edinburgh, pursuing what they called the Strong Programme in the Sociology of Knowledge, and another cluster at the University of Bath, marshaled under the banner of Sociology of Scientific Knowledge, together with a number of other social constructivists outside these schools. As representatives of all of them, Hacking focuses on two well-known books that take a social constructivist stance on science: *Laboratory Life: The Social Construction of Scientific Facts* (1979), by Bruno Latour and Steve Woolgar, and *Constructing Quarks: A Sociological History of Particle Physics* (1984), by Andrew Pickering. Hacking's aim here is not to take sides, but rather to identify just what it is that divides these authors from the scientists that find them infuriating. In fairness, I should mention that I am one of the scientists Hacking quotes in opposition to social constructivism, and it will probably be no surprise that I find Hacking to have been too kind to the constructivists.

On two issues, I have no quarrel with Hacking, or with Pickering or Latour and Woolgar. First, the social constructivists generally do not doubt the *reality* of what scientists study. Pickering does not doubt that quarks exist, and Latour and Woolgar do not doubt that peptides exist. Just as with anorexia and leisure, it is our ideas of these things that are supposed to be socially constructed. Second, many social constructivists (including Pickering, Latour, and Woolgar) have a good understanding of the science about which they write, and are not its enemies.

So where is the argument? Hacking sets out three "sticking points" that separate scientists like myself from social constructivists:

Contingency: Though Pickering does not doubt the existence of quarks, he argues that physics could have evolved in a way that did not take account of their existence. In a sense, this is trivially true: for instance, a collapse of social support for science might have left us endlessly playing with inclined planes and pith balls. The real question is whether we could have continued to make progress at the outer frontiers of physics without taking account of the existence of quarks. I don't think so. I lived through the transition in theoretical physics that Pickering writes about, when the quark idea became generally accepted. Before quarks, the study of strong interactions (the forces that hold the protons and neutrons together inside the atomic nucleus) had gone about as far as it could. We could have gone on studying the showers of particles produced in head-on collisions of other particles, which Pickering offered as an alternative to quark-related experiments, but what would have been the point? We didn't know what theory governed these collisions, and even if we had known, we wouldn't have known how to use it to calculate anything, though not for want of trying. About the only progress that was being made was in exploiting principles of symmetry, like the principle that the equations (whatever they are) governing the strong interactions are left unchanged by interchanging protons and neutrons. But no one knew where these symmetry principles came from. They could not be really fundamental, for they weren't even exact; the masses of protons and neutrons are only approximately equal.

The quark idea had been around since the work of Murray Gell-Mann and George Zweig in the early 1960s, but it met with a good deal of resistance for the obvious reason that no one had ever detected a quark. In my own first work on the unification of weak and electromagnetic forces, I limited myself to particles like electrons and neutrinos that don't feel the strong interactions, because I had no faith in the quark theory or any other existing theory of these interactions. Only later was I willing to bring quarks into the theory, though without much confidence. All this changed with the discovery in 1973 that there is a theory of quarks and strong forces in which quarks cannot in principle be seen in isolation. The thing that convinced me of the truth of this theory and the existence of quarks was that the theory explained the approximate symmetries of the strong interactions in a natural way. I cannot conceive how we could have made this progress or continued to make any other progress in elementary particle physics without the quark idea. The earlier program that Pickering suggests as an alternative—studying phenomena, like head-on collisions of nuclear particles, for which the quark idea is irrelevant—has not been completely abandoned, but nothing much is coming out of it.

"Aha," the social constructivist will say, "it is your idea of progress that is socially constructed!" My idea of progress in fundamental physics is anything that will bring us closer to a simple theory that accounts for all physical phenomena in a natural and unified way. To whatever extent this goal may have been socially constructed in the past, the timbers used in the construction are now largely gone, and the goal stands by itself. Since the time of Newton, the style of physical science has spread among the world's cultures; it has not been changed by the increasing participation of physicists from East Asia or (contrary to the expectations of radical feminists and sexual chauvinists) by the increasing number of women in physics. The goal of a simple unified theory attracts people of all sorts, because we think that the theory is out there to be found, and that we can find it. And the theories we develop in pursuit of this goal are the same theories in all cultures.

Stability: As Hacking points out, one of the most serious sticking points for social constructivist views of science is the remarkable stability of scientific knowledge. Any scientist would like to do something really new, but we do not continually abandon the past in a frantic search for novelty, as some artists do. If (as I think) Maxwell's equations of electrodynamics and the theory of quarks are permanent parts of scientific knowledge that will survive all future revolutions in society, then how can we say that they are socially constructed instead of simply discovered? Hacking quotes disagreements on this point that I have had with the historians M. Norton Wise and Thomas Kuhn, and seems on balance to come down on the side of stability. Hacking suggests that Kuhn may have underestimated the stability of scientific knowledge because he lived in a century that has seen profound changes in physical theory. I would say that physics in this century offers a remarkable example of stability. Each new theory has preserved and even explained its predecessors as valid approximations in appropriate contexts. Kuhn's description of revolutionary scientific change as a "paradigm shift" akin to a religious conversion does not apply to anything in our century's physics, but it does apply to the shift from Aristotelian to Newtonian physics at the birth of modern physical science. It was *this* shift that I think inspired Kuhn's view of scientific revolutions.

Nominalism: This is the doctrine that terms like "pine trees" or "strong interactions" are simply names that we give to freely invented categories. Hacking traces the word "nominalism" back to 1492, but of course the idea is older, going back at least to Abelard. Nominalism stands in opposition to what used to be called realism, the view that there is something out there in objective reality that corresponds to our categories. The modern debate between scientists and social constructivists centers on the reality of scientific facts and theories rather than of categories, but this is not such a big difference; after all, putting things in categories is an elementary version of scientific theorizing. Scientists tend to believe in the reality of their theories, whereas social constructivists tend to doubt this, thinking that scientific theories are only man-made ways of organizing our experiences, like putting ordinary things like trees into categories. Hacking takes no sides, emphasizing how difficult it is to give a precise meaning to a statement that a scientific theory is real. True enough, but it is also difficult to say what we mean when we say that a rock is real. Without claiming to solve such ancient philosophical problems, I would argue that scientific theories share those properties of rocks—stability and independence of societal setting—that lead us to call rocks real.

At one point, Hacking leans over very far to favor the constructivists. He agrees with Latour and Woolgar that one should not say that scientists come to believe in a theory like the Big Bang theory because it is true; rather, he would have us say that scientists came to believe in the Big Bang theory because of observations like the discovery of a uniform background of microwave radiation that seems to be left over from a hot early epoch in the history of the universe. But if we believe in a theory because it agrees with observation, and it agrees with observation because it is true (which is not always the case), then isn't it just a harmless abbreviation to say that we believe in it because it is true? Would Hacking object to the statement that the United States declared war on Japan because of the attack on Pearl Harbor, and insist instead that war was declared because President Roosevelt received reports of the attack on Pearl Harbor?

One last word. I am one of those unfortunate souls who do not enjoy reading most philosophers, from Aristotle and Aquinas to the moderns. I don't believe that it is actually possible to prove anything about most of the things (apart from mathematical logic) that they argue about, and I don't like being pounded by lemmas and syllogisms. Doubtless a weakness of mine, but there it is. Hacking's good humor and easy style make him one of those rare contemporary philosophers (along with Susan Haack, Robert Nozick, and Bernard Williams) I can read with pleasure. But he is still too kind to the Capulets.

Sources

Index

Sources

1. Commencement address at Washington College, Chestertown, Maryland on May 19, 1985; in *Societal Issues, Scientific Viewpoint*, ed. M. Strom (American Institute of Physics, New York, 1987), 202–203.

2. Talk at the Tercentenary Celebration of Newton's *Principia* at the University of Cambridge, June 1987; in *Nature* 330, no. 6147, 433–437 (1987).

3. Talk at the Tercentenary Celebration of Newton's *Principia* at Queen's University, Ontario, October 29, 1987; in *Queen's Quarterly* 95, 96 (1988).

4. *Living Philosophies: The Reflections of Some Eminent Men and Women of our Time*, ed. Clifton Fadiman (Doubleday, New York, 1990), 264–270.

5. Based remarks made at the award of an honorary doctoral degree by the University of Padua, December 7, 1992, in *L'Anno Galileiano* (Edizioni LINT, Trieste, 1995), vol. I, 129–130, and a lecture, "The Heritage of Galileo's Thought in Modern Physics and Cosmology," at the University of Padua, December 6, 1992, in *L'Anno Galileiano* (Edizioni LINT, Trieste, 1995), vol. IV, 381–386.

6. *Twentieth Century Physics*, ed. L. M. Brown, A. Pais, and B. Pippard (Institute of Physics Publishing, Bristol and Philadelphia, 1995), 2033–2040.

7. *Scientific American*, October 1994, 22–27; reprinted in *Life in the Universe* (W. H. Freeman, New York, 1995), 1–9; reprinted as "What Can We Know about the Universe?" in *Scanning the Future: 20 Eminent Thinkers on the World of Tomorrow*, ed. Y. Blumenfeld (Thames & Hudson, London, 1999), 294–303.

8. Remarks at the symposium "What Do the Natural Sciences Know and How Do They Know It?" at the fifth national conference of the National Association of Scholars, Cambridge, Mass., November 11, 1994; in *Academic Questions* 8, 8–13 (1995).

9. Stated Meeting Report at the American Academy of Arts and Sciences, Cambridge, Mass., February 8, 1995; in *Bulletin of the American Academy of Arts and Sciences* 69, no. 3, 51 (1995).

10. *The New York Review of Books,* October 5, 1995, 39–42.

11. *Daedalus* 127, 151–164 (Winter 1998).

12. *The New York Review of Books,* August 8, 1996, 11–15; reprinted in *Science and Social Text* (University of Nebraska Press, Lincoln, 2000).

13. *The New York Review of Books,* October 3, 1996, 55–56.

14. *The New York Review of Books,* June 12 1997, 16–20; reprinted in part in *Lettre International* 38, no. 3, 74 (1997).

15. *The New Republic,* September 8 and 15, 1997, 22–23.

16. *George,* October 1997, 70.

17. *The New York Review of Books,* October 8, 1998, 48–52; reprinted as "Wissensrevolutionen: Paradigmenwechsel: Zur Wissenschaftsgeschichte Thomas Kuhns," *Lettre Internationale* 44, no. 4, 64–67 (1998); reprinted as "La revolución que nunca ocurrió," *Este Pais* 94, 2–9 (January 1999); reprinted as "Une vision corrosive du progress scientifique," *La Recherche* 318, 72–80 (March 1999).

18. *The New York Review of Books,* February 18, 1999, 49.

19. *The Oxford History of the Twentieth Century,* ed. M. Howard and W. R. Louis (Oxford University Press, Oxford, 1998), 22–34.

20. *The New York Review of Books,* October 21, 1999, 46–48; reprinted in German in *Bild der Wissenschaft,* December 1999, 48–49; reprinted in Spanish in *Este Pais—Tendencias y Opiniones* 105, 50–54 (December 1999); reprinted in German in *Das Magazin* 16, 33–36 (April 22–28, 2000); reprinted in Spanish in *Revista de Occidente* 230/231, 158–171 (July-August 2000); reprinted in *The Best American Science Writing, 2000,* ed. J. Gleick (HarperCollins, New York, 2000); reprinted in *The Best American Essays, 2000,* ed. A Lightman (Houghton Mifflin, Boston, 2000).

21. *The New York Review of Books,* January 20, 2000, 64.

22. *The Atlantic Monthly,* January 2000, 107–114; reprinted in Latvian in *Rigas Laiks,* May 2000, 28–35.

23. *The Times Literary Supplement,* February 18, 2000, 8.

Index

About the Author

Steven Weinberg teaches in the physics and astronomy departments of the University of Texas at Austin. His writing for general readers has been honored with the Lewis Thomas Award for "The Scientist as Poet" from Rockefeller University, the Andrew Gemant Award of the American Institute of Physics, and the U.S. Steel Foundation–American Institute of Physics Science Writing Award. For his development of a theory that unifies two of the fundamental forces of nature he received the Nobel Prize in Physics in 1979. President George Bush awarded him the National Medal of Science in 1991 at the White House. He was elected a member of the Royal Society of London, and in the United States he was elected to the National Academy of Sciences, the American Academy of Arts and Sciences, the American Philosophical Society, the Texas Institute of Letters, and the Philosophical Society of Texas. He was educated at Cornell, Copenhagen, and Princeton, and has received honorary degrees from the University of Barcelona, the University of Chicago, the City University of New York, Clark University, Columbia University, Dartmouth College, Knox College, the University of Padua, the University of Rochester, Washington College, the Weizmann Institute, and Yale University. He taught at Columbia, Berkeley, MIT, and Harvard before coming to Texas in 1982.

OTHER BOOKS BY STEVEN WEINBERG

Gravitation and Cosmology: Principles and Applications of the General Theory of Relativity (1972)

The First Three Minutes: A Modern View of the Origin of the Universe (1977)

The Discovery of Subatomic Particles (1983)

Elementary Particles and the Laws of Physics, the 1986 Dirac Memorial Lectures (1987) with R. P. Feynman

Dreams of a Final Theory: The Search for the Fundamental Laws of Nature (1993)

The Quantum Theory of Fields: Volume I: *Foundations* (1995); Volume II: *Modern Applications* (1996); Volume III: *Supersymmetry* (2000)